FORTRESS AMERICA

The Corps of Engineers,
Hampton Roads, and
United States
Coastal Defense

Contents

Illustrations

(following page 70)

Fieldworks around Petersburg during the Civil War
Latrobe's design for Fort Nelson, 1798
Fort Norfolk, 1819
Fort Powhatan at Hood's, 1819
Fort Monroe, Fort Wool (Calhoun), and Hampton Roads
Robert E. Lee's depiction of Fort Monroe, 1832
Fort Monroe, Old Point Comfort, and the Hygeia Hotel
Foundation plan for Fort Calhoun, 1821
Fort Calhoun at its greatest height, 1860
Design for barbette batteries, 1869
Rear of water battery at Fort Monroe, 1868
The 1879 plan for completing Fort Wool
Firing of 12-inch battery on disappearing carriages
Firing of 12-inch seacoast mortars
The 1888 plan for modernizing Fort Monroe
The first Endicott-period plan for Fort Wool, 1888
Emplacement under construction at Fort Monroe, 1890s
Work in progress on mortar batteries, Fort Monroe, 1890s
Fort Monroe from a balloon
Fort Wool, 1984
Rapid-fire guns of Battery Irwin, 1918
Eight-inch rifle on railroad carriage, Fort Monroe
Standard 16-inch emplacement of the 1920s
Submarine Mine Depot under construction, 1940
The ditch at Fort Monroe, 1984

Preface

> The art of fortifying does not consist in rules
> and systems, but solely in good sense and
> experience.—Sébastien Le Prestre de Vauban, ca. 1706

The chief question of American military policy during the 1980s was the "Strategic Defense Initiative." That program proposed to launch into earth orbit a network of satellites to defend the United States from intercontinental missiles carrying nuclear warheads. "Star Wars," said its proponents, would present an absolute, foolproof system of defense and therefore would end forever the threat of war. The country could not afford to do without it, they maintained. The Strategic Defense Initiative had as many opponents as proponents. There is no such thing as an absolute defense, critics said. The technology was beyond our means, the threat was exaggerated, and above all it would cost too much—the country could not afford to have it. Moreover, even if it worked, it soon would be made obsolete by changes in weapons and tactics.

The debate was a very old one in American history. Throughout the nineteenth century and into the twentieth century, military experts portrayed the United States as vulnerable to "sudden war." European countries were ever ready to unleash horrible weapons upon the American continent. At the very least, they would subject us to "humiliation," at the worst to invasion and conquest. Never mind that for over a century the country was at peace with all nations even remotely capable of such evil designs and that its geographical, technological, and economic attributes made the suggestion preposterous. A hostile fleet always lurked over the horizon. Accordingly, from the 1790s into World War II, the United States fortified its coasts and its harbors, seeking safety behind earthen and masonry walls.

The proponents of the Strategic Defense Initiative in the 1980s were physicists, rocketeers, computer designers, and others who represented the nation's most "scientific" experts on military questions. Their predecessors, the experts who promoted earlier systems of defense, were also regarded as the leading military minds of their age. They were the men of the Corps of Engineers of the United States Army.

The Corps of Engineers was founded along with the United States Military Academy at West Point in 1802; it in fact ran the academy for sixty-five years. The corps was established to provide a pool of native talent in military engineering; keeping the regular army at a minimum, the United States stiffened its martial weakness with fortifications. The corps, in other words, was born to build forts and to serve as the brain center of the military establishment. It self-consciously did both for many decades, during which its definition of national defense was the controlling one.

Time passed the corps's ideas by, however, and just as the Star Wars proponents contended with impeccably qualified critics, so after the Civil War the engineers found themselves at odds with other military "experts" who had their own views of defense practices. The corps became hidebound, and ultimately it was shoved aside and then almost entirely out of national defense planning. Nevertheless, for over a century the corps's thinking dominated the defense of the United States against "sudden" war. Its thinking focused on forts.

Fortifications have not received their due in the history of United States military policy. Yet for over a century, coastal fortification was about as much military policy as the country was willing to implement, with the army small and speculations on doctrine being mostly academic. The Corps of Engineers has sparked a large literature, but it is mostly polemical and focuses chiefly on the corps's work in civil engineering. That elides the organization's long-standing focus on military engineering and the important story of its rise and decline as the arbiter of American defense policy.

This work addresses both subjects—fortifications as national defense and the role of the Corps of Engineers in defense planning and implementation. It also goes further, to describe the evolution of the national defense program as represented at Hampton Roads, Virginia. Hampton Roads serves as a case study, and it is

hoped that some attention to how each generation of coastal defenses was designed and built will be useful to the reader. But
there is more. Hampton Roads was the site of the first permanent
English colony and, accordingly, of the first American fortifications. Only a few miles away, French military engineers helped to
defeat the British and win American independence in 1781, making a lasting impression on the leaders of the new nation. In 1813
one of the first American military engineers designed the defeat of
another British invasion at Hampton Roads. Thereafter, the area
was the heart of the country's physical defense. It boasted the
greatest fort ever built in the United States, and it became the
country's major naval center. In the nineteenth century ironclad
warships were introduced there. In the twentieth century Hampton Roads was the hotbed of development for aerial warfare and
antiaircraft defense. Its military engineering history, therefore, is
of more than local importance.

This study is the outgrowth of several investigations over the
years. I should like to thank the many individuals and agencies
who made my research easier, but space is not available to list
them all. With apologies to those whose names are omitted, my
gratitude goes out to Michael Musick, Sara Jackson, James Trum-
bull, John Dwyer, Steve Bern, Robert Cotton, and others of the
National Archives and Records Service; John Slonaker, Richard
Sommers, David Keough, and Dennis Vetock of the United States
Army Military History Institute, Carlisle Barracks, Pennsylvania;
Alice Wickizer and her staff of the Government Documents and
Publications Department, Indiana University Library; Marilyn
Irwin, Lilly Library, Bloomington, Indiana; Karl Kuhlmann, Lucy
Trower, Dorothy Gill, Louanne Pruhs, James Whedbee, Paul
Walker, and Dale Floyd of the Corps of Engineers; Richard Weinert
of the United States Army Training and Doctrine Command, Fort
Monroe, Virginia; and Sarah Olson, William L. Brown III, John
Demer, Roberta Row, and others of the National Park Service.

Lawrence Colborne, comrade of campaigns past, arrived with
ammunition in the nick of time and saved the day. Joseph R.
Blaise, veteran of the Coast Artillery and a survivor of Pearl
Harbor, ventured into no-man's-land and returned with intel-
ligence that sealed the victory, while "Cousin Brent" guarded the
flank. And Mindy Vandeventer, a subaltern of great promise, kept

the troops at their posts and the guns in action in the absence of her captain.

For what is of value in the work that follows, all the foregoing deserve a share of credit. For shortcomings, however, the blame is my own.

Last, as always, thanks to Bea: I could not do it without you.

Know the enemy and know yourself; in a hundred battles you will never be in peril. When you are ignorant of the enemy but know yourself, your chances of winning or losing are equal. If ignorant both of your enemy and of yourself, you are certain in every battle to be in peril.   —Sun Tzu, ca. 400 B.C.

⎍⎍⎍⎍⎍⎍⎍⎍⎍ **1** ⎍⎍⎍⎍⎍⎍⎍⎍⎍

Foundations of American Military Engineering

> All were upon the tiptoe of expectation and
> impatience to see the signal given to open
> the whole line of batteries, which was to be
> the hoisting of the American flag in the ten-
> gun battery. About noon the much-wished-
> for signal went up. I confess I felt a secret
> pride swell my heart when I saw the "star-
> spangled banner" waving majestically in the
> very faces of our implacable adversaries. It
> appeared like an omen of success to our en-
> terprise, and so it proved in reality.
> —Joseph Plumb Martin, 1781

Three ships arrived on the American coast late in April 1607, carrying England's first settlers to the land they called Virginia. They anchored near Cape Henry, which borders the southern entrance to Chesapeake Bay, and an advance party went out to find a place to land. After plumbing shoal waters around Cape Henry, they rowed across a great outlet—Hampton Roads—and tested its northern shore. The results there gave them "good comfort," for which they named "Cape Comfort," soon known as "Point Comfort." From their presence would arise a new nation, and the site of their happy landing someday would become the cornerstone of its national defense.

The colonists camped at Point Comfort for a few days, but for want of fuel and water it was unsuitable as a permanent home. So they ventured up the James River, landing at Jamestown Island, a swampy, pestilential place, on 14 May 1607. When the ships departed a week later, 120 men remained to establish the English presence in North America. Their instructions were simple: build a fort, search for gold and jewels, and discover a passage to the

Pacific Ocean. English-speaking people had staked their claim in America. Now they must defend it.[1]

In selecting Jamestown Island as their home, the colonists settled on an unusually disagreeable spot in a region generally favorable to Englishmen oriented both to the sea and to agriculture. The island is on the north side of the lower James River just above Hampton Roads. The latter is the outlet for the James and other streams, giving ready access to the interior, as does Chesapeake Bay, one of the world's great coastal havens.[2] Escape to the sea and England was therefore easy for the fainthearted, as was the development of a strong economy based on agriculture and waterborne commerce. The region accordingly acquired great and enduring strategic importance. The defense of English-speaking America for centuries required special attention to Hampton Roads and the surrounding Tidewater region. They were the birthplace and schoolyard of American defense policy.

The First Military Challenge

The Jamestown colonists were not the first inhabitants of their new home; Indians had lived in the area for thousands of years. The English intruded upon populous and well-organized peoples, who presented a formidable prospect. Mostly agricultural societies, the Indians of the Tidewater had a fine political and military organization, inhabiting towns fortified by stockades. The white settlers in 1607 encountered a confederacy of Algonquian tribes organized by Wahunsonacock, whom the English called Powhatan. That estimable man had formed an empire in the face of constant threats from the Algonquians' deadly enemies, the Siouan Monacan and Manahoac federations occupying the Piedmont and the upper James, Potomac, and Rappahannock watersheds to the Appalachian Mountains.[3] The presence of a powerful native population—Powhatan's confederacy included over 200 towns, thirty tribes in all, organized for military action—made military engineering important to the colony. The settlers' defensive construction was amateurish, but it met the need of the moment. The first accomplishment was a fort at Jamestown, and as other settlements arose, so did other forts.

The colonists were ambivalent about the Indians, regarding them as either potential trading partners or hostile savages. White explorers moved upriver expecting to meet hostility—and

were surprised when they did not find it. The undisciplined settlers became arrogant, however, and soon wore out their welcome.
The Indians began to push back within two years after Jamestown's founding. By 1622, as far as Powhatan's successor Opechancanough was concerned, affairs had become intolerable. The
Indians fell upon one settlement after another. The whites suffered 350 casualties along with the loss of at least one fort and
several settlements, then counterattacked. There followed a truce
that prevailed until 1644, when Opechancanough tried again. The
final bloodletting led to the Indian leader's own death and the
defeat of the tribes. White power then was too strong for the
natives, among whom diseases took a ghastly toll. The Indians
became a dwindling remnant of their former greatness or retreated westward.[4]

Violent relations between whites and Indians made fortification
essential to community development as the American colonies
grew. The forts built over the next two centuries were usually
crude affairs, characterized by timber or puncheon stockades and
intended for emergencies. They were surrounded by, rather than
surrounding, their communities.

Fort building ran in cycles. The wars of the 1620s and the 1640s
encouraged the whites to look to their stockades. Otherwise, development of strongpoints depended upon the attitudes of those
charged with governing the Virginia colony. In the late 1660s the
government tried to have a series of forts guarding every naviga-
ble river. Fort building and maintenance lapsed, however, as the
Indian danger receded, and the stockades rotted away in the last
two decades of the century. Governor Alexander Spotswood re-
vitalized the program in 1710 during Queen Anne's War, erecting
works to guard the coasts as well as the interior. Another lull
followed until the 1750s, when the French and Indian War revived
the program. Colonel George Washington of the Virginia militia
then took charge of frontier defenses and had over two dozen forts
built or manned in a five-year period. By the time the Revolution
erupted, Virginians—and Americans generally—were fortification conscious.[5]

Guarding Hampton Roads

As the settlers built stockades to forestall Indians, they
also faced threats from without. At various times during the colo-

nial period the Spanish, the French, and the Dutch warred with England, whose colonies required protection. Hampton Roads acquired international strategic significance from the very fact of its settlement by Englishmen.

Queen Elizabeth's government experienced worsening relations with Spain throughout her reign. Englishmen had joined in the exploration of the New World that followed Columbus's voyages but never discovered anything like the riches of Spanish Mexico and South America. Spain was the preeminent world power by the 1560s, dominating maritime commerce. Attempts at peaceful imperial trade ended in a battle between English and Spanish ships at San Juan de Ulua, Mexico, in 1568.

The two nations were bent on war thereafter, as English privateers raided Spanish shipping and ports. Much of the early conflict was a private war between Francis Drake and the Spanish Empire, culminating in 1580 with Drake's return from a raid across the Pacific Ocean. That caused a permanent rupture between the two nations and catalyzed efforts to found an English empire. Drake, Sir Humphrey Gilbert, Richard Hakluyt, and Sir Walter Raleigh, among others, promoted colonization. Their aim was to provide bases for raids on the Spanish, limit Spanish expansionism, spread the English religious gospel, relieve overpopulation of the homeland, and locate treasure.

Gilbert tried and failed to colonize Newfoundland in 1583. The next year Raleigh sent an expedition to the coast of Carolina and Virginia; he established a colony at Roanoke in 1585, but it failed quickly. Raleigh dispatched another colony for the mouth of Chesapeake Bay in 1587. The settlers landed at Roanoke again and disappeared entirely before relief arrived three years later. The "Lost Colony" earned its sobriquet while relations between England and Spain broke into open warfare, culminating in the destruction of the Spanish Armada in 1588. The subsequent waning of conflict caused a brief lapse of interest in Virginia and in colonization in general. It was not until early in the seventeenth century that colonization again became an important concern, with more royal support than it had received during Elizabeth's reign.[6]

The Jamestown colony was founded by private enterprise under a royal charter. For a venture that had strategic as well as economic purposes, it was planned and managed poorly. Over half the original colonists were gentlemen wrapped in the delusion that they would shortly fill their pockets with jewels and go home. About half were dead within a few months, the rest ready to quit. They were sustained by the will of John Smith and by repeated reinforcements. The crown reorganized the parent company in 1609, after which 400 new settlers crossed the ocean. Only 60 remained alive in May 1610, and they were about to give up when more newcomers arrived.

Sir Thomas Dale assumed command the next year and turned the colony into a military operation, imposing stern discipline. His people learned to cultivate tobacco in 1612, gaining a source of income. By 1619 the colony was stable and growing, representative government was established, and the importation of slaves had begun. Within five more years, the company charter had been revoked, Virginia was a crown colony, and its citizens became forceful in asserting their right to govern themselves.[7]

The colony had reason to fear an attack from the Spanish as early as 1609. Jamestown Island, swampy and unhealthy as it was, was not defensible. The settlers determined that the James River could best be defended forward, at its mouth; the site they chose was on Point Comfort. On the northern side of Hampton Roads and the James River, Point Comfort commanded the entrance channel. It was also a nearly isolated peninsula that would be difficult to assault from the land. The colonists built Fort Algernourne there in October 1609. Their forward outpost grew from a simple earthwork to a stockaded strongpoint boasting seven heavy guns and a store of smaller weapons by 1611, when its garrison counted forty men. The position was soon backed by Forts Henry and Charles on the Hampton River. Fort Algernourne burned to the ground in 1612; the fear of Spanish attack had subsided, so it was not rebuilt. Coastal defense suffered general neglect for the next several years.[8]

The Virginia assembly authorized a new fort at Old Point Comfort (as it was now called) in 1629 and appropriated funds the next year; it was completed in 1632. The place had a caretaker, but it was badly decayed by 1639 and fell into ruin thereafter. A Dutch raid in 1667 raised interest in coastal defense again, and the colonists began another fort-building program. A hurricane destroyed a new fort at Old Point before it was completed, international peace was restored, and the customary neglect resumed for several more years. England's increasing involvement in wars early in the eighteenth century revived interest in colonial defense, and by 1711 a number of batteries had been established around Hampton Roads, including one at Old Point Comfort. As had happened before, all such works fell into disrepair as soon as the crisis passed.[9]

The Virginia assembly next addressed the defense of Hampton Roads in 1728. This time, the legislature voted money for a stout work of masonry for Old Point Comfort, named Fort George in 1736 to honor the king. Fort George was one of the few English colonial fortifications that aimed at something more than primitive impermanence. Rather than a lumpy earthwork or rotting stockade, it boasted walls of brick and shell lime. Fort George comprised an inner and an outer wall about 16 feet apart, connected by counterforts 10 to 12 feet apart, making a series of cribs filled with earth or sand. The place looked stronger than it was, however, and deteriorated swiftly. It was not until 1744 that even minor repairs were authorized, and a hurricane, the "Great Gust" of 1749, wrecked the place.[10]

Fort George was to have mounted about a dozen guns, but whether any of them looked out over Hampton Roads is not now apparent. The fort was a fine work for its time and place, but it remained an amateur enterprise. If America's defenses were to achieve permanence, they must involve people who made defensive matters their specialty; that would come later. Meanwhile, the growth of agriculture, commerce, and towns gave Hampton Roads and the Tidewater a strategic importance that became very apparent when the colonies united in rebellion against the English crown.

The Revolutionary War and After

The War for Independence introduced professional military engineering to the Tidewater, but not immediately. The main scenes of action were elsewhere until the war's climax, and they engaged most of the available talent. However, the conflict's last major campaign took place in Tidewater Virginia, with the finest exercise in military engineering to be seen in America for many years—the siege of Yorktown.

The conflict in Virginia began with fratricidal skirmishing, focused on Norfolk, a bustling town on the south side of Hampton

Roads, along the Elizabeth River. About 6,000 people lived there at the end of 1775, their community a prosperous and growing seaport. The colonial uprising forced British governors to decamp for the nearest royal warships. Virginia's governor, John Murray, fourth earl of Dunmore, sought safety at Norfolk, whose harbor hosted a number of British vessels. He found there a community rent by feuding and sometimes open fighting between rebels and loyalists. Dunmore, after a fruitless British attack on a nearby rebel redoubt on 9 December 1775, tried to turn Norfolk into a loyalist stronghold, closing down dissident newspapers and imposing martial law. It was no use; the town was in rebel hands by the first of January 1776.[11]

On New Year's Day, Lord Dunmore ordered the fleet to bombard Norfolk into submission; that only set the place on fire. The conflagration became a holocaust after Virginia and North Carolina militia arrived, set torches to houses belonging to loyalists, then went on a rampage. By 3 January the fires were out of control, and before long seven-eighths of Norfolk burned to the ground. The place was thoroughly ruined, its people moving elsewhere to avoid recurrent British occupation and raids. Dunmore's heavy-handed attempt to secure Norfolk for the crown gave the port its first work of military engineering. British forces erected a stockade, called Fort Murray, on the Eastern Branch of the Elizabeth River. It was supposed to cover the land route to Norfolk, but it was abandoned in the face of the rebel attack.[12]

Dunmore was driven from his last base on Gwynn Island on 9 July 1776 by a rebel assault supported by fire from works erected by Engineer John Stadler of the Virginia forces. The area enjoyed relative quiet for two years. As the British effort to restore royal control intensified, however, Virginia felt the effects economically and hardened its resolve. On 15 May 1778 the governor and Council offered a number of military recommendations to the General Assembly, including: "For the further Security of Trade four small Batteries are ordered to be erected on the Eastern Shore. These seemed absolutely necessary as the Enemy are closely blocking up the Channels through which our Vessels formerly passed into North Carolina, and our Trade must in future be principally carried on by the Way of that Shore."[13]

The recommendations did not produce an organized system of defenses. They did reflect apprehension about the region's vulnerability, fears aggravated by a growing British presence on the Virginia coast. Norfolk and Portsmouth (opposite Norfolk on the Elizabeth River), because of their fine harbor, became the focus for British and rebel forces alike. Militiamen erected their first work of defense by early 1779. Fort Nelson was named for Brigadier General Thomas Nelson, signer of the Declaration of Independence, veteran of the Continental army, then commander of Virginia's state militia, and in 1781 Thomas Jefferson's successor as governor. Located just west of Portsmouth, Fort Nelson was an earthen parapet 14 feet high and 15 feet thick on the river side. Armed with heavy guns and abundantly provided with powder, the place seemed secure enough to guard Portsmouth's wartime commerce.[14]

On 10 May 1779 British naval forces under Admiral Sir George Collier and land forces under Major General Edward Mathew launched a combined assault against the Norfolk-Portsmouth area. As Mathew reported, Fort Nelson was no obstacle: "About Three in the Afternoon the Army was landed at the Glebe, on the Western Shore of Elizabeth River, just out of Cannon Shot of the Fort. As the Troops landed, the Column moved to invest the Fort. The Enemy, perceiving that their Retreat would be cut off, evacuated before we could reach the South Branch of Elizabeth River." Portsmouth was open to British depredation. The royal forces took or destroyed everything, including 137 vessels, causing over two million pounds in losses to the Americans. Among the objects of British attention was Fort Nelson. It may have been defenseless, but Fort Nelson was stout. According to Mathew, "The Engineer has been employed for many Days, with near one Hundred Blacks, to destroy the Fort, which was so substantially constructed, as to give us a great Deal of Trouble in the Demolition."[15]

Fort Nelson was not the Norfolk-Portsmouth area's only asset. Knowing that he had taken one of the finer harbors on the Atlantic Coast, Admiral Collier pleaded for permission to extend his raid into a permanent occupation:

> This Port of Portsmouth is an exceeding safe and secure Asylum for Ships against an Enemy, and is not to be forced even by great Superiority. The Marine Yard is large and extremely convenient, having a considerable Stock of seasoned Timber, besides great Quantities of other Stores. From

these Considerations, joined to many others, I am firmly of Opinion, that it is a Measure most essential for his Majesty's Service that this Port should remain in our Hands since it appears to me of more real Consequence and Advantage than any other the Crown now possesses in America; for by securing this, the whole Trade of the Chesapeake is at an End, and consequently the Sinews of the Rebellion destroyed.

General Sir Henry Clinton wanted Mathew and Collier on hand farther north, however. Delayed only by the time it took to demolish Fort Nelson, the British departed on 24 May 1779.[16]

The British returned repeatedly to intimidate the rebels. Governor Jefferson decided in 1780 that, having laid waste the principal seaports, the British would carry their raids into Virginia's interior. Under his orders, the state was to build a battery at Hood's, on the James River below Richmond. The work was not completed when Benedict Arnold launched a raid from Hampton Roads in January 1781. Hood's was outflanked and abandoned by its garrison, and the enemy spiked its guns and moved on to Richmond.

Baron von Steuben, inspector general of the Continental army, was in Virginia to organize defenses against the raids of Arnold and General William Phillips and generally to improve the state's military condition. He had with him Colonel John Christian Senf, formerly chief engineer of South Carolina's military forces. Steuben and Senf persuaded Jefferson late in January 1781 that a new battery was needed at Hood's. Senf designed and supervised the project, but once again the state was slow to act. When a second British raiding party arrived in April, the place was unfinished and unmanned. It was finally completed during the summer, after its need had passed.[17]

The floundering of the Americans was exceeded only by that of the British. By midsummer, Arnold and Phillips were under the control of General Lord Cornwallis, who not only roamed without apparent purpose but lost communication with Clinton in New York. When American forces under the marquis de Lafayette threatened his rear, Cornwallis first maneuvered against him. The British commander then retired to Yorktown, which he reached on 2 August 1781 with orders to hold open a port for the British fleet. George Washington, commander in chief of the Continental army, and the comte de Rochambeau, leading the allied French forces, realized that Cornwallis could be trapped. As the French navy approached Chesapeake Bay in mid-August, the allied armies left New York and hastened to cut the British off. The French drove away the British fleet in the Battle of the Virginia Capes on 5 September, and on 14 September Washington and Rochambeau joined Lafayette at Williamsburg. Two weeks later the American and French forces laid siege to Yorktown, and the war became an exercise in military engineering.[18]

There were about a dozen professional military engineers in the allied forces around Yorktown and considerably fewer among the British. The latter laid out and built a perimeter of earthworks to protect Cornwallis's force until it was evacuated by the navy. The French naval victory off the Virginia Capes made that impossible, and Washington and Rochambeau then planned a methodical effort to force a British surrender.

Chief among the engineers in charge of the allied program were Washington's chief engineer, Major General Louis Duportail, and Lieutenant Colonel Jean Baptiste de Gouvion, who mapped Washington's route of march. Both were officers of the Continental army, Frenchmen with training and experience. Rochambeau let the investment of Yorktown be an American operation, assisted by the French, secure in the knowledge that it would follow French principles. Duportail and Gouvion plotted the siege according to the geometrical methods of the great French military engineer of the seventeenth century, Sébastien Le Prestre de Vauban. In his system, the design and construction of earthworks were the province of the engineers, while the siege itself belonged to the artillery. The allied engineers, therefore, must design works that not only would invest the enemy but would be useful for and protective to heavy cannons and their infantry defenders.

As some troops probed the British positions, others went into the surrounding woodlands to fashion gabions—wood and brush baskets to be filled with earth and form the body of the fieldworks—and such objects as fascines—bundles of sticks used especially to fill in the enemy's ditches during infantry assaults. Cornwallis abandoned his outer positions and pulled into the immediate area of the town, and the siege began on 6 October. The first work was a ditch and an earthen parapet paralleling the enemy position about 600 yards away. By 11 October, the troops had dug a zigzag, or approach trench, and begun a second parallel about 300 yards forward. After troops and sappers overran two British redoubts, the second parallel extended swiftly almost to the York River, providing a series of heavy gun positions. The siege then became the job of the artillerists, who hammered Cornwallis's troops mercilessly. The British surrendered on 19 October, and the independence of the United States had been won.[19]

The work of Duportail and his colleagues introduced professional engineering into the American military scene. They also embodied a new presence in American public affairs—the Corps of Engineers of the Continental army, ancestor to the Corps of Engineers of the United States Army. According to its own tradition, the origins of the Corps of Engineers may be traced to the Battle of Bunker Hill, 17 June 1775; Colonel Richard Gridley, the Continental army's first chief engineer, designed the fortifications there and was wounded in the fighting. Actually, on the day before the battle the Continental Congress authorized one chief engineer and two assistants for the "Grand Army" and another chief engineer and two assistants "in a separate department." On 27 May 1778 the lawmakers authorized three companies of "sappers and miners," who became part of the Corps of Engineers formed on 11 March 1779. The enlisted men of the corps were comparable to those of other Continental army units, but owing to a shortage of professional engineers in America, most of the officers were European. Their contributions were crucial, and their success at Yorktown promised that engineers would be part of American military affairs in the future.[20] The Corps of Engineers left Virginia along with the rest of the army after Cornwallis's surrender, and after two years at West Point in New York, it disappeared during the near abolition of the military establishment. The end of the war brought peace to America and a desire to rebuild commerce.

Commerce did revive, at Hampton Roads and elsewhere. In fiscal year 1790, Virginia was first among the thirteen states in duties paid on tonnage of vessels entering the United States. In addition, Virginia was second in the value of its exports and first in exports of domestic produce, tobacco being the most important.[21] Of all Virginia ports, Norfolk with its fine harbor and strategic location benefited most from the growth in commerce. In 1793 and 1794 several thousand French refugees from the slave revolt in

Santo Domingo sought safety there. The first contingent arrived in a fleet of 137 ships escorted by eight French warships in July 1793. According to one of the newcomers, Norfolk had about 500 houses, mostly frame, "with wonderful attics and lightning rods," and a construction boom was in progress. The public square was surrounded by shops and taverns, the biggest building in town was a mercantile establishment, and the waterfront sported a line of large warehouses. "The harbor is large enough to accommodate 300 ships," said the Frenchman. "The river channel is from 120 to 140 *toises* [ca. 750–900 feet] wide, and at ordinary high tide its depth immediately in front of Norfolk is about 18 feet." Despite a climate that was oppressively hot and "quite deadly," Norfolk was booming.[22]

By 1800 Norfolk had more than regained its population of 1775 and had removed the last scars of its destruction in 1776. Commerce thrived, and ships crowded the harbor. Similar prosperity blessed nearby Portsmouth, which also entered national service as a naval base. The Gosport shipyard there had been established in 1767 by Scotsman Andrew Sproule; because he became a loyalist, the state took possession of the property at his death in 1776 and bought it in 1780. Gosport established itself as the builder of galleys and other vessels for the Virginia navy and of some ships for the Continental navy. British Admiral Collier wrecked the place after the fall of Fort Nelson, and there was not much left when the city of Portsmouth annexed it in 1784. Shipbuilding resumed in 1794 and 1795, and before long the Gosport shipyard had produced the sloop-of-war *Chesapeake* as its first work for the federal government. On 15 June 1801 Gosport became a United States naval shipyard.[23]

The Revolution and its aftermath made Hampton Roads and environs vital to the economy of the emerging United States. The accidents of geography that supported the commerce also made the area a natural military and naval center. The war had demonstrated that defense was a common, national problem, better addressed by the central government than by the states individually. It also had shown that warfare was not properly an amateur enterprise when directed against a competent and determined enemy. Postwar prosperity demonstrated that American well-being depended on commerce. The nation soon decided that its commerce must be defended by the navy, but both commerce and the navy required secure bases. Their protection could be achieved by military engineering—a field in which the French, as at York-town, had proved themselves very impressive. Their example was not forgotten when the United States faced its first crisis in coastal defense—nor, indeed, for generations to follow.

⎍⎍⎍⎍⎍⎍⎍ 2 ⎍⎍⎍⎍⎍⎍⎍

America's First Two Fortification Systems

It is deeply to be lamented, that a very pre-
cious period of leisure was not improved towards forming among ourselves engineers
and artillerists; and that, owing to this neglect, we are in danger of being overtaken by
war, without competent characters of these
descriptions. To form them suddenly is impossible. Much previous study and experiment are essential. If possible to avoid it, a
war ought not to find us wholly unprovided.
—Secretary of War James McHenry, 1798

The United States assumed it could do without a national military force after the Revolution, except for a few dozen
men guarding stores; the states would defend the nation with
their militias. There seemed nothing to fear from abroad, so there
was little attention to defense of the seacoast. Slightly more concern focused on the interior, where several states had extensive
claims, eventually transferred to the federal government.

Worsening relations with Indians and British agents north of
the Ohio River brought military questions to the fore. In 1785
Congress called out 700 militiamen for frontier service. In the
years following, others saw service against domestic disturbances.
Such forces were ineffective, and the need for a better military was
among the considerations that led to the adoption of the Constitu-
tion. On 7 August 1789 Congress established a Department of
War, and a month later it authorized a regiment of 700 men.
Secretary of War Henry Knox soon called for a standing army of
2,033 officers and men. A "small corps of well-disciplined and well-
informed artillerists and engineers," he said, "and a legion for the
protection of the frontiers and magazines and arsenals" were all
the United States needed, if its militia was efficient. Congress took

no action on Knox's plan. The army remained small, scattered along the frontier, and short of specialists.[1]

World events, however, would not leave the new nation free to prosper without feeling some anxiety about its physical security. A recurrent sense of national danger would force the United States to adopt some sort of military policy and program more palpable than a small body of men or occasional drafts of militia to cow Indians and guard supplies. But the nation had a constitutional fear of standing armies born in the history of English civil wars and the pre-Revolutionary experience. The militia, therefore, would answer when danger threatened.

The difficulty with militia was that it was ineffective against a competent and professional enemy army, and the nation's leaders knew this. A small professional army might stiffen the spine of the amateur forces, but that would require some form of professional education for its officers; in 1802 the government established a military academy. More generally, however, early American leaders remembered the example of Yorktown in 1781. There, competent engineers applying European principles had used structural methods to accomplish what had eluded the American army—the defeat of British regulars.

Fortifications thereby presented themselves as an answer to the national defense problem. Competent engineers could secure the country's assets without the dangers and expense of a large professional army, because their works would provide a safe rallying point for militia or volunteers. Instead of a large army, therefore, the United States needed only a modest cadre of technical experts, and these it provided by establishing the Corps of Engineers. Engineers would fortify the vulnerable parts of the coast; and in the absence of a sizable army, coastal fortification became the centerpiece of American military policy for over a century.

The First Emergency in Coastal Defense

The regular army's defeat of the Indians at the Battle of Fallen Timbers, Ohio, on 20 August 1794, and the contemporaneous suppression of the Whiskey Rebellion in western Pennsylvania by militia forces insured that the country would have a small regular army for continuing service but would call on the militia for emergencies. The need for a more refined arrangement was demonstrated by another crisis in 1794. The revolutionary

French were at war with Britain and other European powers, and the United States, also a revolutionary nation, was caught in the middle. Surprisingly, however, the first foreign military threat to the United States came from France.

Hampton Roads felt the effects early. A French agent, Moreau de Saint-Mery, explained its situation in the spring of 1794, as he observed the rebirth of Fort Nelson: "The seizure of American ships going to or leaving French ports has given Congress the idea of taking some measures to defend the coast. As a consequence, fortifications were begun in Norfolk in May 1794; this particular fort is built of earth and located on the left bank of the Elizabeth River about a Thousand *toises* below Portsmouth. In it the militia has already mounted a twenty-four hour guard composed, in all, of 14 men. The building of this fortification is under the direction of M. Rivardi, an Italian engineer."[2]

Norfolk hosted armed vessels from both France and Britain. "The case of the ships of war on opposite sides being at Norfolk," Moreau observed, "creates a delicate and dangerous situation for that town, which nothing can remedy so effectually as its being put in state of respectable defence. Indeed some information has been received by which it would appear that the Doedalus had fired a shot in a very unjustifiable manner." Actually, on the night of 8 May, American soldiers fired three shots at the English frigate *Daedalus,* which was "anchored so close within range that a cannon ball was found in one of the ship's water casks." That "reprehensible act," Moreau claimed, proved that Norfolk's citizens favored the French over the British. When the same happened with a French ship. Moreau reported, the American responsible was executed.[3]

What was happening at Norfolk did not reflect sympathy with either Britain or France, however. Early in the year, Congress formed a Committee on Fortifications to develop a plan to prepare the nation's ports for a possible war with France, during which French raids might be expected. More generally, the lawmakers wanted to provide secure bases for the navy, whose creation was to be authorized shortly. One of the first naval bases was to be on the Elizabeth River, putting Norfolk high on the list of harbors requiring protection. The committee offered a plan to build works at sixteen "ports and harbors of the United States as require to be put in a state of defence." Part of the plan called for twenty-four guns at Norfolk—batteries, embrasures, and platforms; a redoubt, with embrasures; a magazine; a blockhouse or barracks; and contingencies, all itemized with a total cost of $3,737.58.

Norfolk would be garrisoned by one subaltern, one sergeant, a corporal, two musicians, and seventeen privates. All forts would be manned by infantry; "it is, however, supposed, that some of the artillery officers in service might be used on the present occasion, and that part of the infantry officers might be chosen for the purpose, who would soon acquire a tolerable degree of knowledge in the use of cannon." The highest officers in the program would be two majors, "to act as inspectors to be constantly employed in visiting the posts."[4]

In other words, although the lawmakers desired preparedness, they wanted above all to spend as little money as possible, whether on construction or on artillery specialists. When Congress made the first funds available in March, Secretary Knox faced the monstrous challenge of implementing a scheme designed in excessive detail by penny-pinching amateurs. He must rely mainly on the states to carry out the program, because they were supposed to provide much of the armament and to secure the land for the forts. Moreover, the army lacked fort-building experts, and the actual construction was a responsibility the states refused. Knox hired foreign engineers resident in the United States and sent them out with broad instructions, the main one being that their works were to be simple and cheap. For political reasons, the engineers must seek the approval of state governors at every stage of planning.[5]

One of the immigrants was John Jacob Ulrich Rivardi, a Swiss living in Pennsylvania whose military engineering background came to Knox's attention. The secretary of war appointed Rivardi a "temporary Engineer in the service of the United States," cautioning him that he was entitled to no military rank or privileges. The engineer, however, was known throughout 1794 as Major Rivardi. In February 1795 he became a major in the Corps of Artillerists and Engineers, a rank he held until his discharge in 1802.[6] "The Bearer, Major Rivardi," Secretary Knox wrote the governor of Virginia on 28 March 1794, "is the Gentleman whom the President of the United States has appointed for the purposes of fortifying Baltimore, Alexandria and Norfolk. The circumstances of the latter Port being considered as the most pressing, the Engineer will at present make but a short stay at Baltimore or

Alexandria. He is directed to furnish your Excellency with a copy of his instructions." Rivardi's instructions—like those to all other engineers—told him to go to the three cities immediately and, if the governors of the states were nearby, to tell them what his instructions were, inform them of his movements, and let them see the results of his studies. Knox's orders also presented an abstract of what Rivardi was to build and budgets for Baltimore and Norfolk but not for Alexandria, which had been inserted into the program without an estimate.[7]

Knox emphasized the low cost estimates for Rivardi's projects. The forts must be earthworks, he said; timber was to be used only where good earth could not be had. The parapets should be sloped properly, sodded inside and out, and sown with seed of "Knot grass." Embrasures should be framed with joists and faced with 2-inch planks. If the batteries were isolated from towns, they should be enclosed in redoubts. The garrisons could live in barracks or blockhouses, "as shall be judged most expedient. But, in general, as the garrisons will be weak in numbers, a Block-house mounting one or two small pieces of cannon in its upper story, will be more secure, and therefore to be preferred." Blockhouses, Knox demanded, should not be designed for heavy cannons or more than fifty men; tents could house any other men assigned to the work.

As for the redoubts, or main works, each ought to hold 500 men, "and perhaps the idea ought to be embraced in the first instance that they would be of such extent as to admit timber casemates to be erected hereafter, so as to enable the Garrison to resist in some tolerable degree a bombardment." But Rivardi was not to build any casemates at present except for powder magazines. The latter were supposed to be ventilated and dry, each to hold 150 rounds for every cannon in a fort. Knox also advised Rivardi to use his careful judgment on the siting of the magazines, on protecting the works by "freezers" (fraises) or palisades, and on giving the redoubts embrasures or having their guns fire "en barbette" (over the parapet). Knox preferred parapets without embrasures.

Rivardi also was to exercise his judgment on the choice of sites for the forts, "under the direction of the Governor." He was enjoined further to find a good supervisor for each project. Supervisors were to be empowered to hire workers and materials, "but every thing must be previously estimated and calculated by you,"

Knox told the engineer. Finally, Rivardi would receive four dollars per day in compensation and reimbursement for any reasonable extra expenses for which he had receipts. He was to provide the secretary of war and the governors copies of all plans, surveys, and other productions and to report his progress weekly. Rivardi should go to Norfolk from Baltimore as quickly as possible.

Rivardi wrote to Virginia's Governor Henry ("Light-Horse Harry") Lee from Baltimore on 3 April 1794, promising to leave for Norfolk soon. He wrote again on 15 April, emphasizing his interest in getting the work started quickly and asking early approval of his preliminary plans. Lee had wanted to meet Rivardi at Alexandria, but the engineer said that he wanted to defer going there until Norfolk was "in a state of defense." In any event, Rivardi said, he was eager to meet the governor.[8]

Rivardi reached Norfolk early in May. Soil tests revealed that his structures would require foundations of brick or at least fascines. The engineer wrote to the president on 6 May, telling him that he preferred brick and asking for an extra $1,200. Nothing, he said, should be left undone for the safety of the country. "At my arrival here," Rivardi reported, "I found an extensive coast, requiring multiplied points of defence; a soil, loose, without the least adhesive quality; the people, though disposed to assist with all their power, much less numerous than at Baltimore; and I compared, with great concern, the sums allowed for the latter place and Norfolk." Rivardi was relieved to report that he had found an ally in Lee: "Since my arrival, public money has been expended only for the purchasing of materials and tools, the labor being done almost entirely by the public. Governor Lee's exertions and indefatigability have removed almost every obstacle which I undoubtedly should have experienced had he not been here as soon as myself."[9]

Knox informed Lee on 9 May that Rivardi would remain at Norfolk and that someone else would do the work at Alexandria. Ammunition, carriages, and implements were about to be sent to Norfolk, and nine 18-pounder guns were also on their way. Recruiting and officer appointments were about to begin for new artillery units, "some of which will repair as soon as possible to Norfolk. In the mean time Captain Hannah will be ordered to Norfolk with twenty Recruits, which he has at Alexandria." A month later, Knox told Lee to assume that Rivardi's extra $1,200 would be forthcoming. Knox also remarked, "I shall also be particularly obliged by a confidential opinion of the merits of Mr. Rivardi as an Engineer, and, as far as you have opportunity, of his talents as an artillerist."[10]

Rivardi was a candidate for appointment to the Corps of Artillerists and Engineers about to be authorized. His performance, viewed from afar, was less than impressive. Congress appropriated more money for fortifications on 9 June, and the War Department revised the budget for all projects. Knox was plainly concerned about whether Rivardi could prosecute an expanded workload. Rivardi's difficulties were not of his own making. Bad weather retarded the work and gave him severe chest pains. Moreover, on 9 June he was unable to assemble any workers. He sent Lee a plan for Fort Nelson that day and promised plans for "the distillery" and "the necessary drawings of Crany Island & its defences" in short order, with copies to go to the War Department on 10 June.[11]

Rivardi had arrived in Norfolk with the understanding that Fort Nelson would be the main work. Before long, however, he perceived that a more comprehensive defensive system would be necessary. That merely added to his difficulties, and to Lee's concerns. Governor Lee was heavily involved in the fortification of the Norfolk area by early June and visited the place to see for himself the troubles besetting the engineer. Rivardi continued to have difficulty finding workers, and the local militia commander was no help. In addition, the engineer was burdened by jealousies between Portsmouth and Norfolk over the splitting of appropriations between Fort Nelson and other locations. Some individuals began to carp about "the difficulty of manning so many different places." Nevertheless, Rivardi assured the governor that "I stop my ears and proceed as fast as circumstances will allow." Most of the ground before four batteries was removed by 15 June, assembly of fascines was "going on briskly," and the engineer was about to survey Craney Island downstream from Fort Nelson and finish plans for its defense. He also reported progress on a map of the Elizabeth River. Finally, "The brick work and all the traverses which were begun are compleat, and the whole of my attention will be next week directed towards the fitting of the batteries."[12]

Thanks to Rivardi's energy and the increased funding, Fort Nelson and the other defenses of the Norfolk area became more substantial than first conceived. Fort Nelson was now something more than a simple earthwork, and it was matched by Fort Norfolk across the river and masked by works on Craney Island and batteries at other locations. Rivardi sent Knox new plans for Fort Nelson and Fort Norfolk on 24 June 1794, saying that the latter place was already coming along nicely. Promising further plans for Craney Island and a chart of the river, Rivardi said that Lee was pressing him to start work at Craney Island. He would do so shortly because Fort Norfolk was so far along.[13]

Rivardi's progress was short-lived, however. He told the War Department on 6 July that he had forwarded all of his plans, but that bad weather, "deficiency of cash," and the departure of citizens to tend crops "have put a temporary slowness in every thing here." He filled his time surveying and mapping the Elizabeth River and was nearly finished with a map of Cape Henry and all river mouths in the region. Rivardi's plans for Craney Island met resistance from the War Department, but he replied to Knox that it was "a spot of too much importance to be neglected, as all the vessels are forced to come under point blank shot of it." Rivardi asked for clearer instructions on construction of shot furnaces and for more expense money. Last, the engineer complained again that labor was short and that he had not heard from the secretary of war since the middle of June.[14]

Rivardi's circumstances improved by 11 July, when he sent Lee a copy of his map of the Elizabeth River. He explained that his plans to chart a larger area must be deferred because of the travel involved. In a more positive vein, he continued: "Notwithstanding the bad weather, Fort Nelson goes on tolerably well. Most of the platforms are placed, the new lines more than half raised towards the land side, and the powder magazine so advanced that if the weather is more favorable it will be completed in a few days." Rivardi was unable to do any work at Fort Norfolk as long as he was busy at Fort Nelson. Some of the area's cannons had arrived, however, and the engineer had mounted them where they commanded the harbor entrance.[15]

In one week at Fort Nelson, 7–13 July 1794, Rivardi's crews removed earth, made fascines, moved and placed bricks, moved and laid lime, laid fascines and platforms, and generally brought the work along. Rivardi expected the place to be done by the end of July. Things were not so happy at Fort Norfolk, however, and the engineer was delighted to learn on 19 July that Lee would pay another visit. He hoped the governor's presence might end "those vile little cabalings so detrimental to public good." Whatever the local politics, Fort Norfolk was held up more by an absence of money than anything else. Rivardi intended to move his resources to Norfolk when work on Fort Nelson concluded; "by that time, I hope we shall have a pittance from the Country and every thing will go better."

Rivardi meanwhile had concluded that Old Point Comfort should have been the place to start work on Norfolk's defense. Its absence from the program allowed him only to assert that Craney Island could be defended by just 100 men if it was matched by works at Old Point. His recently completed plans for Craney Island showed eleven guns and the cheapest design he thought practical. Rivardi asked Lee to approve his Craney Island plans immediately so that he could start planning construction. Most important, he concluded his long letter of 19 July, he had not received money promised by the War Department, "which hurts us much."[16]

On 20 July Rivardi wrote a long letter to Secretary Knox, explaining his lack of progress. He had expected good things, he said, from public contributions,

> for I thought that, in a small country, where public welfare ought to be the chief aim of every individual, no jealousy, no parties, could be found. I do not think, however, that there exists, any where else, such ridiculous divisions as here. The inhabitants of Portsmouth expected all the means should be employed in protecting their side, and refuse their assistance at Fort Norfolk. The inhabitants of Norfolk, not to be behind hand, refuse their's at Portsmouth, and will not even do any thing on this side of the river, if every one does not work in turn. The consequences are obvious: Money must purchase all the labor, and, even then, it is sometimes impossible to get more than thirty or forty hands. This is not the only inconvenience to be complained of. There is a large number of dissatisfied men who object altogether to fortification, from the

same principle for which they object to every measure of Government. Some would rather *bush fight*, (as they call it) in case of a war, and the fact is, I fancy, that they had rather not fight at all. I drop this disagreeable subject: the only thing is to be deaf, and do what the safety requires.

Rivardi reminded Knox that his budget was $4,937, including the $1,200 added during the spring, but he had received only $2,700. Nevertheless, he claimed that "a few hundred dollars more" would finish the project. Also, he had not yet received $1,500 reportedly sent to him to pay for mounting the cannons. There were now to be thirty guns guarding the Elizabeth River, rather than the original twenty-four, and the engineer believed that his budget should be raised accordingly. Rivardi also forwarded his final plans for Craney Island, along with further argument for construction at Old Point Comfort, but he seemed resigned that the War Department would not sanction work there.[17]

The federal government was trying to push ahead with the fortification program at all locations. Knox told all the engineers on 24 July to provide information on the "quantity of ground" required for their forts to advise federal agents who would purchase the land. He said in closing, "Permit me to urge the season of the year, which is advancing, as a strong inducement to placing all the fortifications under your direction in a state of defence, and of completing them, as far as possible, with the funds which have been designated." Rivardi was not the only engineer behind schedule.[18]

The War Department pressed ahead on land acquisition, soon reverting to the original plan for the states to buy the properties, though now with federal money. Governor Lee meanwhile urged Knox to keep the works at Norfolk underway, and on 30 July the secretary informed the governor that funds destined for Norfolk, not counting weapons but including purchase of land, totaled $6,737.52. The Treasury Department, he assured Lee, would disburse the money as quickly as it was needed. Quickly was a matter of interpretation, for land acquisition as for all else. Fort Nelson stood on state property. Fort Norfolk, however, was on private land, and the legitimacy of its occupation was dubious. On 17 December 1794 Treasury Secretary Alexander Hamilton authorized the governor to purchase the property, "provided the cost does not exceed one Thousand dollars." Finally, on 21 May 1795, at a price of $972, a deed was secured to 4.31 acres "for defense of Norfolk and Elizabeth River."[19]

Work on the forts accelerated late in 1794, but money remained a problem. Advising Congress early in December that the forts "are in considerable forwardness," Knox submitted a revised total estimate of $225,500. He blamed the threefold increase over the first plan on labor costs. The House Committee on Fortifications, amid heightened international tensions, proposed that up to $500,000 in new money be granted to complete the program, along with $100,000 a year to maintain the forts. On 28 January 1795, tensions reduced, the committee reversed itself, saying that $46,500, including $3,000 for Norfolk and Portsmouth, would complete the program, and recommended only $50,000 in new appropriations. The latter sum appeared on 3 March 1795.[20]

The stern economy that governed the fortifications program based the defense of the Elizabeth River upon two unimposing earthworks, Forts Nelson and Norfolk. Only the most incidental development took place at Craney Island, and none at Old Point Comfort. What Rivardi accomplished by December 1794, however, was impressive, considering the lack of local cooperation. He had almost but not quite completed the job.[21] Politics interrupted the fortification program. Disagreements with the French waned early in 1795, and with them the government's interest in pursuing the work. United States troops occupied Norfolk's defenses in small number for a brief while, then were ordered elsewhere. They were to be replaced by Virginia militiamen, who were not eager to assume the responsibility.[22]

The protection of the Norfolk area was not complete. Congress trifled with the idea of finishing the job in 1796 and sought an estimate from Secretary of War Timothy Pickering. He replied with a description of Rivardi's accomplishments and a $5,000 estimate.[23] Except for some stabilization measures and works at New York City, the House Committee on Fortifications recommended in May 1796 against spending more money on the fort program. The First System of American seacoast defenses, as it would later be called, remained as it was. A few works, including Fort Norfolk, had small garrisons that gave them some maintenance, and others, including Fort Nelson, were substantial enough to retain their basic integrity. The majority, mostly simple earthworks or "star forts" (from their shape), melted swiftly into the landscape. Early in 1797 the War Department asked for money to hire four engineers and repair several of the forts, including both at Norfolk. Although Congress gave the department $24,000, it spent only $3,000 by early summer, when the secretary asked for $200,000 to complete all works, including $10,000 "to improve and complete the works at Norfolk." His request fell on deaf ears.[24]

A Corps of Engineers at Last

Secretary of War James McHenry found a more attentive Congress during 1798, when relations with France worsened again. The Jay Treaty of 1794 symbolized to the French a pending alliance between America and the British, although it was nothing of the kind. The French overreacted and threatened American shipping. A delegation sent to Paris late in 1796 was at first rebuffed, then in 1797 met with an insulting demand for loans and a hefty bribe for three French agents identified as "X, Y, and Z." The XYZ affair led Federalists to demand preparations for war: "Millions for defense, but not one cent for tribute," one of them put it. From March to July 1798 Congress passed at least twenty laws to strengthen national defenses. Among the new developments was the United States Navy, only tentatively initiated earlier, increases in the regular army, a "provisional" army that took form only on paper, and an expanded fortification program.

The martial impulse soon moderated. There was a general belief that the navy ought to provide the first line of defense, with the army relegated to coastal protection, and in the fall of 1798 the British navy destroyed the French fleet. Republicans also feared the designs of the Federalists and what they might do with a large professional army. Finally, by 1800 relations with France calmed. The most positive result of the episode was a modest start on a new fortification program.[25]

The war scare also eventually resulted in one enduring legacy in the army organization—the United States Military Academy. The need for professional military education had been recognized for decades. Most of the leaders of the Continental army had advocated a national academy for regulars, militia, or both. They had also pointed out the need to correct the nation's shortage of professional engineers. The fortification program of 1794 made the lack of native engineers painfully obvious; proficient artillerists were an equally serious need. Congress sought to solve both problems in 1794 by authorizing a Corps of Artillerists and Engineers, assuming that the two subjects were aspects of the same art. The conjunction of the two professions, compounded by jealousies between them, did not meet the need for a pool of native engineers. "A power to procure from abroad one distinguished engineer, and also an officer of artillery, and suitable appointments for the same," was something that Secretary McHenry requested in 1798, along with power "to provide for the appointment of an inspector of fortifications."

Congress wanted to expand the army as it was, not embark on new ventures in organization. The passing of the crisis provided an opportunity to consider weaknesses in the American system. The country could not, after all, hire French engineers every time it threatened to go to war. The Republicans under President Thomas Jefferson made changes in the military organization after 1800. Jefferson was determined to reduce the army drastically, but he also wanted it to be useful. A proponent of education, he wanted to improve the officer corps. The military legislation of 1802 cut the size of the regular army and also established the Corps of Engineers and the United States Military Academy (the two were at first identical). Before too many years passed, engineers trained at West Point, the academy's home, made their mark all over the American landscape.[26]

The Corps of Engineers long remembered its origins as the brainstem of the army. It controlled West Point until 1867 and thereby stamped its own outlook on the education of the officer corps. For decades the cadets were taught that engineers had a higher calling, as theirs was the more exacting and "scientific" branch of the military art. The Corps of Engineers, accordingly, took the top graduates of almost every class, leaving the rest to the fighting branches. The best military brains, said the implicit message of West Point, were to be devoted to engineering; lesser minds could patrol the frontier and chase Indians, which is what most of the army did for most of the nineteenth century.

Engineering in the army meant above all fortifications. The self-appointed elitism of the Corps of Engineers had both negative and positive aspects over the years. Negatively, it discounted the status of the officers of the fighting arms, who attained good experience in small-unit leadership but not much preparation for modern warfare. The engineers' dominance also meant that theirs was the controlling definition of national defense policy; for decades more attention was paid to forts than to the relative roles of forts and the army in national defense. Third, engineering in the army tended to be rigorously academic, tied as it was to higher education at West Point. The engineers were highly schooled in textbooks and lectures plotting well-established (that is, European) principles, but except on the battlefield nineteenth-century army engineers were less than adept at matching theory to real-world conditions. Technology, therefore, repeatedly passed the fortifications by, and each time the engineers found it hard to adjust. Ultimately, their inflexibility cost them their control of both West Point and defense planning.

There was a positive side, however. The engineers gave all officers a good, technical education and integrated essential engineering knowledge into the military art. Moreover, the army generally in the nineteenth century was very small, its strength and brainpower dissipated in service as a frontier constabulary. The Corps of Engineers remained focused outward, however, and through fortification work was oriented always toward the army's first mission—national defense—and to modern warfare. When major wars did develop in 1846 and 1861, and when the army began to modernize late in the nineteenth century, that legacy of the Corps of Engineers benefited the organization as a whole.

The corps began to evidence its special character from the outset. The first chief engineer and superintendent of the military academy was Colonel Jonathan Williams, nephew of Benjamin Franklin and enthusiast for the military sciences. Williams's passion was fortifications. He built several before his resignation in 1812 and established design principles characteristic of the so-called Second System of American coastal forts—the "castle" of tiered casemates propounded by the Frenchman Marc René, marquis de Montalembert, and, most notably, the curved facade that survives at Fort Norfolk. During his tenure the defenses of Hampton Roads were transformed once again.[27]

The Waning of the First System

Interest in reviving the fortifications program preceded the establishment of the Corps of Engineers. The Committee on Fortifications of the House of Representatives declaimed the "alarmingly exposed" condition of American ports and commerce in March 1798. The committeemen also observed that Congress had appropriated $115,000 for fort construction in June 1797, of which the War Department had spent only about $17,000. Secretary McHenry asked for more money to complete the forts but equally was adamant about competent manpower to garrison them. Militiamen, he said, were useless in the fortifications. Before long, he had authority for more forts and men as well.[28]

The War Department became more active during 1798, showing special attention to the Norfolk area. Benjamin Henry Latrobe, who among other accomplishments supervised construction of the United States Capitol, was appointed to survey Norfolk's defenses at the height of the war scare. He believed Fort Norfolk to be of little real or potential value and regarded Fort Nelson as the area's strongpoint. Latrobe redesigned Fort Nelson according to the latest European principles. The most important change was the elimination of the unfinished embrasures in favor of continuous parapets, which is how Knox had wanted the forts to appear originally. Latrobe further improved Rivardi's design by adding bastions, a ravelin, and a wider ditch and by completing the traverse shielding the barracks and magazine. When Fort Nelson was reconstructed in the following years, it was according to the great architect's plans.[29]

Activities like Latrobe's did not result in immediate reconstruction of the nation's harbor defenses. The waning of the war scare, combined with reliance on the navy, kept appropriations low. "The fortifications erected for the defence of our cities and harbors," Secretary McHenry warned in 1800, "cannot yet be considered competent to afford this security. Many new and extensive works, even at those places where the fortifications are advanced, will yet be required to render any of them a secure asylum for our navy."[30]

McHenry asked for $100,000 to work on forts at several Atlantic Coast ports, including Norfolk. However, interest in the program declined with the rise of the Republicans. McHenry's successor, Henry Dearborn, decided to see things for himself, starting with the Norfolk area. During a visit there in 1802, he ordered Fort Norfolk to be abandoned as useless and a new fort erected at Ferry Point in what is now Berkeley. Protests from the people of Norfolk scuttled his plans at Ferry Point, but nothing could stop him from abandoning official interest in Fort Norfolk. Fort Nelson became the Norfolk area's strongpoint, with a garrison of about seventy officers and men after 1802. Beginning that year, officers resumed work on Fort Nelson according to Latrobe's plan; they spent about $15,000 before 1806. The president identified Fort Nelson as Norfolk's only fortification in 1806 and asked for additional money to complete it. "Considerable improvements and repairs [are] yet necessary," he said.[31]

Dearborn reported to Congress at the end of 1806 that a number of coastal fortifications had been improved during the year. Among them was Fort Nelson, where twenty heavy cannons had been mounted. That fort was still not complete, a distinction it shared with most of the country's harbor defenses. A special Senate committee declared in December that Fort Nelson was among a considerable number of forts requiring appropriations for completion or repair.[32] It was one thing to recognize the country's need for fortifications; however, positive action required incentive. That the British soon provided.

The Second System of Fortifications

Relations with France and Britain became further strained as the Napoleonic wars continued. The focus of most disputes was on American shipping, which ignored the blockades of either side. The British, moreover, claimed the right to search American ships for deserters from their own vessels. The British frigate *Leopard* raised the dispute to a new height on 22 June 1807 when it stopped a frigate of the United States Navy, the *Chesapeake,* off Hampton Roads. The British commander claimed that four men aboard the American vessel were British deserters and demanded their surrender. The American commander refused, and the *Leopard* opened fire, killing three men and wounding eighteen others. The British then boarded the ship and removed the alleged deserters, and the *Chesapeake* limped into Norfolk. The country was outraged by the news. Affairs worsened when Britain stepped up its campaign in October. The Jefferson administration, wanting to avoid war, retaliated in December with the Embargo Act, interdicting foreign trade with American ports.

Virginians reacted more violently to the *Chesapeake* affair than did the rest of the country. Mobs at Hampton destroyed casks of water stored there for the British navy. The citizens of Portsmouth and Norfolk wore crepe for ten days. Governor William H. Cabell started to strengthen the state's defenses and so notified the president. Cabell rushed arms and supplies to Norfolk and told the militia commander there to muster his men. The entire state mobilized.

The Royal Navy closed in. The British commander challenged Norfolk's right to terminate intercourse with his ships and early in July threatened to blockade the region, moving into Hampton Roads. Cabell declared that move an act of war, predicted an attack, and on 6 July ordered much of the state militia to Norfolk. He informed the president that he regarded his state as having been invaded, and he demanded federal assistance. The state then began to fortify all of Virginia's coast, sent more troops to Norfolk and Hampton, and put the rest of the militia on alert. Secretary of War Dearborn sent supplies to Virginia and planned to raise 10,000 militiamen around the country.

The first emergency passed as suddenly as it had arisen. On 15 July, Cabell reported that the British ships had left Hampton Roads, and he canceled further militia movements. Those already called up were disbanded and returned home by mid-August. But Norfolk remained aroused. A public meeting on 29 June resolved to support the repair of Fort Norfolk and called on citizens to offer their own labor and that of their slaves. Ship masters using the harbor also called for refurbishment of the area's defenses and offered assistance. President Jefferson accepted the latter offer in July.[33]

As the people of Norfolk did their best with Fort Norfolk, the federal government again examined its fortification program. A new project for fifty places (not all actually fortified, as it happened) was prepared, and over the next five years the United States spent over $3 million to build its Second System of coast defenses. It was the first major public work to be carried out in part by engineers who were American in birth and training—the early products of West Point and the Corps of Engineers.[34]

Fort Nelson was the only federal work at Norfolk when the program began, Fort Norfolk being a local project. "Considerable improvements are now progressing," Secretary Dearborn told Congress early in December. In another message he explained other necessary measures: "Fort Nelson is on a commanding site, and in good repair, but may require some additional support in the rear. A battery at Portsmouth, one on Hospital Point, and one on the site of the town, for the cover of heavy cannon on travelling carriages, aided by a suitable number of gunboats, are thought capable of affording a competent defence against any naval armament that can approach the town. . . . In addition to the above contemplated works, it will probably be considered expedient to erect a strong enclosed work at or near the mouth of Lynnhaven creek."[35]

The War Department also addressed the defense of rivers. A list of ports and harbors of minor importance included three in Virginia. On the James River, said Dearborn, "at a place called Hood's Point (considered a commanding position) a strong battery, covered by a redoubt, may, perhaps, when aided by the floating force below, be considered a sufficient defence for the towns above." On the York and Rappahannock rivers, "for each a battery and block house."[36]

Construction started in 1808, spurred on by the return of British warships to Hampton Roads; the federal government spent over $1 million on forts that year. Between 1808 and 1812, at least five parcels of land and timber came into federal ownership to secure sites at Hood's and in the Norfolk area, including a site for a battery at Ferry Point. Fort Nelson benefited first; it was essentially completed during the year, with liberal use of masonry. Fort Norfolk also enjoyed professional expertise and federal money when it returned to federal service in 1808. The old earthwork gave way to an irregular enclosed work of brick masonry, whose main feature was a semielliptical battery defended by irregular bastions on flanks and rear.[37]

The president reported that the government had purchased a site and materials for a "strong battery at Hospital Point, near Norfolk . . . but too late for completing it the present season." At Hood's, the new Fort Powhatan was taking form as "a strong battery of mason work . . . and a regular enclosed work, on an eminence commanding the battery." Jefferson described the work there as "in considerable forwardness, with barracks for two hundred men, nearly completed, including quarters for the officers." Resting on the site of the 1781 battery at Hood's, the new fort was a square parapeted masonry work with sharp salients on three walls and a water battery. It was designed for twenty-five guns, all of which were mounted by 1814, but the place did not see its first action until the Civil War.[38]

The federal government spent another $36,800 in the Norfolk area during 1809, before the emergency quieted. At the termination of the work, Secretary Dearborn described the monuments the engineers had left at Norfolk: *"Fort Nelson,* on the west side of the entrance of the harbor; an enclosed work of brick and earth, defended by half bastions, calculated for forty guns, thirty-three mounted; with a brick magazine and barracks for two companies, including officers. *Fort Norfolk,* on the opposite side; an enclosed work of masonry, calculated for thirty guns, ten mounted; with a brick magazine, and barracks for two companies, including officers."[39]

That was the conclusion on the Second System of fortifications at Hampton Roads. Little changed at either Fort Nelson or Fort Norfolk before the War of 1812 except the number of guns and the rate of neglect. Fort Powhatan remained about half finished, not even armed until the war began. Nor had anyone involved shown Rivardi's appreciation of Craney Island and Old Point Comfort.[40]

The Corps of Engineers was not heavily involved in the Hampton Roads defenses during 1808 and 1809, because the organization was too small and young to take complete charge everywhere. Fort Nelson was the fruit of Latrobe's genius, while the general placement of fortifications reflected Rivardi's efforts years earlier. The first attempts at rehabilitating Fort Norfolk rested in the hands of amateurs. The corps nevertheless played a part. Fort Powhatan was a carefully designed complex that reflected more than passing familiarity with military engineering. More apparent was the touch of Jonathan Williams's teaching that was demonstrated in the elliptical facade of Fort Norfolk's main battery. Today, Fort Norfolk is the best surviving example of a Second System fort with a curved masonry face. As such, it graphically illustrates the state of the Corps of Engineers' art early in its history.[41]

Hampton Roads at War Again

The fort builders aimed not at posterity but at immediate requirements. Forts Nelson and Norfolk had not been put to the test, nor had the uncompleted Fort Powhatan and batteries around the Norfolk area. Relations with Great Britain continued ominous, and in 1812 another war began. When it was over, the army's engineers would ask themselves whether they had built well enough—whether they had assembled a true system of defense or merely so many forts and batteries.

Walker Keith Armistead answered the second question before the war hardly began. He was the Corps of Engineers' second-ranking officer, a lieutenant colonel. It was a reflection of Norfolk harbor's importance that so high an officer from so small an organization should be assigned to its defense from 1812 to 1814. Armistead graduated first in the class of 1803 and enjoyed a rapid rise to the rank of major in 1810. He became a lieutenant colonel in July 1812 and in 1818 chief of engineers. Jonathan Williams sent him to Norfolk.[42]

Armistead had two tasks. One was to put existing defensive works in the best condition possible. The other was to develop a general scheme to defend the entire area, integrating established positions with any new ones that appeared advisable. He had few resources to work with and was probably the senior among the few regular army officers around Hampton Roads. He had mostly militia for manpower. Armistead therefore had to plan works to be built by forces who were uncertain commodities as workers and to be defended by those same people, equally uncertain as soldiers.

The British navy arrived in Hampton Roads early in February 1813 and proclaimed a blockade of the southern coast of the United States. An attack could be expected at any time, so the community was galvanized into activity, with citizens, militiamen, and seamen offering their services. Fort Nelson was in fairly good condition, well armed, and ready for action. Fort Norfolk was also well armed and respectable on the water front. It was nearly defenseless from the rear, however, so during the spring the militia built a series of earthworks to guard it.

Armistead perceived that reliance upon the two forts could be a mistake. Each was easily outflanked, as had been the first Fort Nelson of the Revolution. Neither Armistead not anyone else seriously considered defending Hampton Roads from the most forward position, Old Point Comfort. The resources for developing works there were lacking, and the British navy already commanded Hampton Roads. Armistead nevertheless believed that Norfolk should be defended from an advanced position, so that it would not have to receive an attack at all. The obvious place to anchor such a defense was on Craney Island, at the mouth of the Elizabeth River. Armistead evidently drafted a plan and sent it to the War Department in the spring. On 10 June 1813 Secretary of War John Armstrong told Congress, "A work on Craney Island, for the better protection of Norfolk, has been commenced, and is mentioned here but from a belief that the importance of the position will be found to justify an extension of the present plan."[43]

Armistead had no chance to complete the works on Craney Island, as the British fleet moved west of that area on 21 June and prepared to attack. The island now had a fort on its east side and redoubts on the west covering the land approaches, sporting two 24-pounders, one 18-pounder, and four 6-pounders manned by navy men, militia, and some troops from Fort Norfolk. Virginia's militia general Robert B. Taylor was in command and during the night rushed forces to the island, about 750 men in all. Militia and boatmen guarded the east flank and the river entrance with a line of twenty gunboats anchored in the channel while the troops on the island set up simple breastworks among the uncompleted fortifications.

The British landed about 2,500 marines and seamen on the mainland, intending to storm Craney Island and move on to Fort Nelson. The Americans stood their ground, cannons driving the enemy back. Meanwhile, about fifty barges loaded with British troops approached the north end of the island, where withering fire repulsed them with the loss of five boats. The Battle of Craney Island was over in a few hours, and the astonished British had lost about 200 men.

The British consoled themselves by landing at Hampton and brutally sacking the town. The raid served no strategic purpose, but it frightened the people in Richmond, who called out their militia and formed a vigilance committee to look to the capital's defense. "The confusion is past and we are safe not only for the present but against any enterprise which the enemy may meditate," the Richmond *Enquirer* reported on 2 July. "Several hundreds of men are ready at Fort Powhatan to breast the first shock—and at this moment there cannot be less than 4,000 men at Malvern Hills, Sandy Point and their immediate vicinity—exclusive of the troops in Richmond, etc." The committee of vigilance was not so sanguine, however, and decided that building defensive works would be a waste of time. In the committee's opinion, Richmond was geographically indefensible.[44] The committee's advice was either to declare Richmond an open city in the face of a British invasion or to take to the woods to fight. That was little solace to people plagued by rumors of a British advance from Hampton. The alarm subsided quickly, however, and by the middle of August most of the militiamen had gone home. The whole region nevertheless felt the effects of the continuing British blockade of Hampton Roads.[45]

In February 1814 the vigilance committee petitioned the governor to have Fort Powhatan "put in a proper situation for defending the passage up the James River." The governor passed the request to the legislature, which in turn referred it to the national authorities. Except for the mounting of guns, however, no important work was done at Fort Powhatan before the end of the war. Richmond became alarmed again when British invaders appeared in the Chesapeake Bay in July, and by September the town was almost empty, as its citizens fled an expected invasion. When Baltimore and Washington proved to be the targets, they returned to await the end of the war.[46]

The War of 1812 underscored several principles for American defense in the future. One—renewing a lesson from the Revolution—was that haphazard and amateurish fort building and militia musters did not answer the need for a continuing and competent defensive system. Another legacy was a growing appreciation of the strategic significance of Hampton Roads and the vulnerability of Chesapeake Bay and the American interior—including the national capital—to invasion or blockade if Hampton Roads was not proof against attack. Among the nation's growing body of defense experts, the officers of the Corps of Engineers, the war demonstrated that harbors and the country as a whole would best be defended from forward positions, to prevent rather than resist attack. Academically trained and chastened by experience, the engineers established themselves permanently at Hampton Roads in 1816 and set about perfecting a Third System of fortifications. It became their collective mission to insure that the United States should never again suffer invasion—or even insult—from an enemy fleet.

⎍⎍⎍⎍⎍⎍⎍⎍ **3** ⎍⎍⎍⎍⎍⎍⎍⎍

The Golden Age of Coastal Fortifications

> If War ended in a single solution, or a number
> of simultaneous ones, then naturally all the
> preparations for the same would have a ten-
> dency to the extreme, for an omission could
> not in any way be repaired. . . . But if the re-
> sult is made up from several successive acts,
> then naturally that which precedes with all
> its phases may be taken as a measure for
> that which will follow, and in this manner the
> world of reality . . . takes the place of the ab-
> stract, and thus modifies the effort towards
> the extreme.   —Carl von Clausewitz, ca. 1824

The federal capital was in ruins, epitomizing the national embarrassment that had been the War of 1812. The enemy
had invaded the country several times, twice from the sea. American ground troops seldom did well. At Baltimore, however, a fort
held back the foe. "It is not necessary to make of the coast a
fortified line," a French artillerist advised after the war. "No other
plan, it is true, will prevent descents upon an extended coast, but
it [is] impossible to protect every point. . . . All that can be done is
to guard the most important positions."[1] With such guidance, the
United States again looked to its forts.

With fort building representing the nation's determination to
defend itself, it was inevitable that the fort builders—the men of
West Point and the Corps of Engineers—would play a major role in
military affairs. The corps and the academy were young and not
yet in control of the army, but they found themselves in a congenial
climate at the end of the war. The army had new leaders—Winfield Scott, Jacob Brown, Alexander Macomb, Andrew Jackson,
and others. They were not graduates of West Point, but they were
vigorous, forceful, and talented and had succeeded where their

elders had failed. Most important, they were fierce advocates of military professionalism, and in that West Point's—and by extension the Corps of Engineers'—promoters. In addition, Macomb was oriented to engineering; he became chief engineer in 1821 before rising to commanding general.

The corps was still youthful and accordingly full of the arrogance of the young and well-educated. It had reason collectively to believe that it held the key to national defense: its works at Baltimore had saved that port, and the Battle of Craney Island had shown what one capable engineer could do in the face of great danger.

The War of 1812 left a lasting impression on the Corps of Engineers, serving as it did as the corps's baptism by fire. In decades to come the organization repeatedly evoked memories of British invasion to support its own plans and had a ready answer to anyone who thought an enemy invasion unlikely—it had happened. In fortifying the American coast against another such attack, the corps inevitably gave special notice to a locality where an engineer had vindicated engineering as the foundation of national defense—Hampton Roads, where the British came to grief against the well-planned defense of Craney Island.

In the process, the engineers were inclined to look with awe at the sources of their technical principles—the accumulated wisdom of Europe, especially France, where military engineering over generations had been carried to a high art. Europe might be the enemy, but it was also still the teacher.

The Third System

The American government, for the last time, hired a Frenchman to guide its military development. He was Simon Bernard, an engineer under Napoleon who entered the United States with recommendations from the marquis de Lafayette. President James Madison appointed him assistant engineer 16 November 1816, with the status of a brigadier general. Bernard became head of the Board of Engineers for Fortifications the same day. The appointment of a foreigner to direct American defense policy offended the pride of the Corps of Engineers, or at least that of some of its leading lights. Chief Engineer Joseph G. Swift objected to Bernard's elevation on grounds of military security, as well as personal and professional insult; he resigned his commission in

1818. Fortifications board member William McRee felt the same way and left the army in 1819.[2]

Besides Bernard, the original board included Major William McRee and Major Joseph G. Totten of the Corps of Engineers and Captain J. D. Elliott of the United States Navy. When considering the defense of any place, they added to their number the senior local officer of the engineers. With two exceptions, the board's membership changed often. Bernard remained until 1831. Totten worked on fortifications all his life and became a world leader in the subject. He enormously influenced coastal defense policy.[3]

The Board of Engineers' mission was to plan a defense of the seaboard. That resulted in the so-called Third System of American defense, the works of the 1790s becoming known as the First System and those of 1808–12 the Second System.[4] The board's plan was the first that was truly systematic, however. The board visited vulnerable or strategic points and developed principles for their defense. The panel started issuing interim reports annually in 1818, its first full set of recommendations in 1821, and a revised package in 1826. As early as 1818 work was underway at a number of places, including Hampton Roads, and the War Department had decided that all coastal fortifications would be "permanent works" (that is, of masonry) large enough to defend their positions. According to Chief Engineer Swift, "This mode of construction is the best calculated to secure the object for which fortifications are erected upon the extreme seacoast of the Union, namely, to protect the important and valuable points. It is also true economy. The expense once incurred upon the above principle, will not require to be repeated."[5] This assertion not only implied that maintenance of the new forts would not be necessary but rested on the assumption that the highly geometrical tactics and technology guiding both the design of and attacks on forts would never change.

Secretary of War John C. Calhoun wanted to increase the Corps of Engineers to twenty-three officers to supervise fort construction. Legislation setting the size of the army in 1821 gave him twenty-two officers, supplemented by ten in the Topographical Bureau—an increase over the previous twenty-two in all.[6] Having authorized enough engineers, Congress next demanded the War Department's defensive plans. That brought forth the Board of Engineers' first full report on 12 February 1821. The United States, said the board, required (1) a strong navy; (2) fortifications at important ports and harbors; (3) improved interior communications; and (4) the regular army and the militia. Taken all together, the board promised, those constituted a true "system" of national defense.

The navy would require more than ships. The board proposed two "naval arsenals," one at Burwell's Bay in the James River and the other at Charlestown, Massachusetts. In addition, Hampton Roads, Virginia, and Boston Roads, Massachusetts, were to become the navy's "great rendezvous," with Narragansett Bay, Rhode Island, an "accessory." Several lesser stations and "ports of refuge," along with facilities on the Gulf of Mexico, would complete the navy's place in the system. Fortifications would protect the navy's bases. The board outlined three classes of forts, the first to hold 2,440 men in peacetime and 20,305 "in case of siege"; the second 1,030 and 8,615; the third 1,120 and 9,042. The total cost was estimated at $17.8 million, while the troops required to man the works would number 4,690 in peacetime and 37,962 in war, "supposing them, which is beyond all probability, all besieged at once."

The board went on to argue that with a small peacetime army, the country must assemble militia and volunteer forces rapidly in an emergency. Improved land and water communications, therefore, were a military necessity. As for regulation of the regular army and militia, the board's plan included more posts than the 6,000-man army could garrison while meeting its other responsibilities.

The four-part defensive "system" proposed by Bernard and his colleagues was an elaborate combination of military idealism and political realism. It was already an article of faith that the navy would provide the country's "first line of defense," protecting commerce abroad and operating against invaders near shore. The army therefore claimed its share of the national defense budget by safeguarding the navy's bases. Moreover, with the events of the last war fresh in mind, the board answered public fears of another British invasion. In the tactics and technology of the day, an invader must seize a suitable port to sustain both his naval and land forces. Strong harbor defenses would prevent that.

Realistically, the board eschewed a direct challenge to the nation's persistent assumption that militia—generally recognized as a failure—and volunteers were an effective substitute for a profes-

sional army in wartime. The board coyly referred to a need for a
"change in policy," which really meant a larger army and the
upgrading of the militia into a competent reserve force; Bernard
and his partners may have believed that such a change would
become generally desirable on its own account. In the meantime,
fortifications were offered as the most practical way to make militia forces equal to a professional adversary—as putatively had
happened during previous British invasions.

Last, in calling for improved roads and canals, the board touched
a nationally resonant cord. Except for the older coastal cities, every
community (and its congressman) wanted better access to the
world of commerce. Making transportation a military necessity not
only paved the way for the engineers to grow into a formidable
public works force; it insured that old coastal centers and newer
interior provinces both would find cause to support the engineers'
activities. To get support for forts desired by maritime ports, their
congressmen must appreciate the strategic value of the canals or
river improvements desired by interior states, and vice versa.
Whether or not the board really thought the potential through to
its logical conclusion, it laid the foundation for pork-barrel politics
and bureaucratic growth and made it likely that these impulses
could be employed to benefit the interests of the military bu-
reaucracy. It was simply a matter of time before Congress would
realize the benefits to its members of spending the unprecedented
sums of money that the board's vision would require.

Unprecedented they were. Bernard did not believe his ideas to
be unrealistic. He did expect their expense to meet opposition,
however. Completing the Third System would take many years, he
said, but appropriations ought to be sufficient to do the work as
rapidly as possible. He recalled the case of France, which took fifty
years to fortify but more than once had been saved by its forts. It
was cheaper, he continued, to build forts than to defend places
without them. "The defense of our maritime frontier by permanent
fortifications, and even the expense of erecting these fortifications,
will thus be a real and positive economy," he averred. Fortifications reduced the number of points of attack and therefore the
number of men required to guard the seacoast. Moreover, they
allowed diversion of resources to the navy, which could hold invaders away from shore.[7]

Secretary Calhoun won suitable financing for a few years. New forts began to rise at ports on the coast, and in 1823 the secretary urged Congress to appropriate $100,000 annually for ten years to buy cannons for them. Calhoun's successor, James Barbour, reported in 1825 that the new fortifications were coming along well. The secretary temporarily dissolved the Board of Engineers in March, and until the fall of the year its members separately inspected works in progress. Now down to two members, Bernard and Totten, the board issued its final report in 1826; it was mostly the same as the 1821 version, with a few minor adjustments. Secretary Peter B. Porter reported in 1828 that so many forts were complete or nearly so that he had moved about half the training companies from the Artillery School of Practice at Fort Monroe to other posts. Still the work continued, and in 1831 Secretary John H. Eaton called for an increase in the Corps of Engineers because many projects lacked a supervisor or assistant. Instead, he faced a decrease in fort appropriations, which hit zero in 1834. Work stopped altogether in 1835.[8]

The Third System was almost two decades old. The Senate wanted a review of the program and on 18 February 1836 asked the War and Navy Departments to advise on "appropriations that would be necessary to place the land and naval defences of the country on a proper footing of strength and respectability." The two departments replied through the president on 8 April.[9]

Secretary of War Lewis Cass reviewed the United States' geographic relation to its potential enemies and avowed "that our first and best fortification is the navy." The first task was to increase the navy, so that it could fight an enemy at sea, keep him from our shores, and protect our commerce. Cass thought it unlikely that any large invasion force would land in the United States; it certainly would be impossible to conquer the United States. As in the past, invaders would control only their own neighborhood. The country lacked the city centers that dominated European nations and was so diffuse that not even the loss of the capital in 1814 had threatened the nation.

Cass concluded that European fortification systems would be useless in America. "I consider some of the existing and projected works larger than are now necessary, and calculated for exigencies we ought not, with the prospects before us, to anticipate." Moreover, the forts wasted the manpower and money required to maintain them. "Certainly the stronger a work is the less it will be exposed to danger; but this would not furnish a sufficient reason for making its defences out of reasonable proportion to its exposure." The Board of Engineers had planned on massive invasions and protracted sieges of isolated works; Cass thought naval raids more likely than invasion in the event of war.

"To apply these remarks to the plan of fortifications partly completed," the secretary began with Fort Monroe at Hampton Roads. That place, he observed, covered sixty-three acres and required 2,700 men to man its projected 412 guns in time of war. The object of Fort Monroe and nearby Fort Calhoun was to keep the enemy out of Hampton Roads. That was important, but Hampton Roads was not the only good anchorage on that part of the coast. Furthermore, its forts did not close off Chesapeake Bay, and "the occlusion of this roadstead does not secure Norfolk, important as it is from its commerce and navy yard." Clearly, there were limits to the effectiveness of any fort in that area; something smaller than Fort Monroe could do the job.

Every port worth attacking ought to be defended, Cass said, but the ruling criteria should be two: "1. That [the forts] be able to resist any naval batteries that will probably be placed against them; and 2. That they be also able to resist any *coup de main* or escalade which might be attempted by land." Building forts to resist long sieges was absurd. The present plan, therefore, was overdrawn, especially considering the increases in American population, industry, and transportation since 1815. "Certainly some of the facilities and means of defence are augmented beyond any rational expectations," he said. Cass believed that steam-powered floating batteries were the wave of the future in coast defense, "and of the most important advantages to be anticipated from the works at Old Point Comfort is the security they will afford to the floating batteries cooperating with them, and which will find a secure shelter in Hampton Roads. A hostile fleet, about to enter the Chesapeake, would certainly calculate the means of annoyance to which it would be exposed by these formidable vessels."

Cass noted that two fortifications bills were pending in Congress. One would continue works already begun, with a large appropriation. Cass approved, but he suggested that many of the projects could be reduced or eliminated. The other bill would appropriate money for nineteen new forts and $600,000 for steam batteries. The secretary said that few of the new forts were called for and that the steam batteries should be funded first at $100,000. Cass in closing asked for an increase in the Corps of Engineers, the appropriation of a project's entire cost at its start, and the development of armories and arsenals.[10]

Cass's report was accompanied by a contradictory statement from Acting Chief Engineer Joseph G. Totten, who saw no reason to revise the plans of the Board of Engineers, even after so many years. Totten believed that forts should be built to withstand the worst that could be brought against them. As for Hampton Roads, Totten would not change earlier thinking. He clearly had an emotional investment in the work of the board and enduring faith in Simon Bernard. Former Chief Engineer Swift recalled that Bernard stamped his imprint onto the fort program over the better judgment of some of the other board members: "On this board McRee and myself found Bernard rather shy in giving his reasons for the preference of any part of the plan that was his own. . . . His uniform reply was 'Gentlemen, your plan is very good, *mais,* I prefer my idea.' We both said we had a right to his reasons in the spirit of his employment. McRee and myself also preferred a smaller enceinte to the work at Old Point. I had so stated to Mr. Secretary Calhoun, but we deferred to Bernard's preference and popularity, and yet we did not receive his reason for so large an enclosure."[11]

Fort Monroe at Hampton Roads was Bernard's grandest creation, but it was too much for its time and place. The fort embarrassed the War Department, because Cass's objection to its scale was picked up by critics of the construction program. The House of Representatives raised the issue again in 1840. Totten, then chief engineer, admitted that the design was a mistake but defended Bernard and by implication his own concurrence in Bernard's grand vision. "But it has long been too late to remedy the evil," he suggested, attributing the error to Bernard's European background. Moreover:

> The mistake is one relating to magnitude, however, not to strength. Magnitude in fortifications is often a measure of strength; but not always, nor in this instance. Fort Monroe might have been as strong as it is now against a water attack, or an assault, or a siege, with one-third its present capacity, and perhaps at no more than half its cost. We do not think

this work too strong for its position, nor too heavily armed; and as the force of the garrison will depend mainly on the extent of the armament, the error has caused an excess in the first outlay chiefly, but will not involve much useless expense after completion.[12]

Hampton Roads was a popular resort, so Fort Monroe made vacationing congressmen wonder whether other parts of the Third System might be equally extravagant. The result was a dearth of appropriations for some years. None came forth in 1835, while one for 1836 could not be spent because of its lateness and the shortage of engineers, now heavily involved in river and harbor improvements. Appropriations were delayed again in 1838 while Secretary of War Joel R. Poinsett reassured Congress that completing the Third System would not require a larger army.[13]

The questions raised by Cass and members of Congress never redefined the country's fortification needs. Instead, there ensued a protracted struggle over funding. There was some money available in 1839 but not enough to suit Totten. Some appropriations were granted in 1840, but not until July, and the Treasury made them unavailable for the whole year. Totten almost begged for reform, because the engineers lost half the work season year after year. Meanwhile, the House of Representatives asked for a report on the coast defenses. Totten responded with an updated version of the old Board of Engineers reports and another defense of Fort Monroe and Simon Bernard. Now he had an ally in Secretary of War John C. Spenser, whose defense of the original plan was modified only by the addition of steam floating batteries. "As the system cannot be abandoned," avowed the secretary, "what has been commenced should be completed." Contradictions persisted. The House Committee on Naval Affairs asserted in 1841 that American harbors were vulnerable to French or British invasions from the West Indies, but no one else took the threat seriously. In 1842 troops were moved to several coastal forts, but construction stopped again that year.[14]

Another issue then complicated the program. The assignment of garrisons revealed that few forts had anywhere for the men to live except casemates, which were little more than man-made caves. In 1820 the War Department had ordered the engineers to build accommodations at the permanent fortifications. The barracks and quarters at Fort Monroe were among the few erected by the Corps of Engineers, however. The engineers thought such appurtenances beyond their concern and in some cases objected that they would interfere with fields of fire or compromise military appearances. In the mid-1820s barracks construction at permanent forts was assigned to the Quartermaster Department, which built no such housing while the forts were unoccupied. In 1843 the army's leaders fell into dispute over the housing issue. Commanding General Winfield Scott complained of an "extreme want" of barracks and hospitals and objected to the fact that "cramped and most unwholesome *casemates* now [were] in general use for both purposes." "[I]t would seem against the interest of the country and the credit of the Government," he averred, "to lodge troops, with their sick . . . in such miserable places." The surgeon general also was outraged, and the quartermaster general rebuked the engineers.[15]

Secretary of War William Wilkins spoke up in 1844, asking for money for decent accommodations. Totten, never having lived in a casemate, maintained that they were adequate habitations. He compromised only so far as to suggest that housing be provided at a distance from the forts. Wilkins made him change his mind, and through 1851 the Corps of Engineers offered estimates for housing. No separate barracks money appeared before the Civil War, and Totten let the issue drop. When the War Department raised it anew in the 1850s, however, housing was designed into new forts started during that decade.[16]

Work on the coastal forts resumed in 1844. The next year the engineers worked on forty-eight forts, forty of which were then declared ready or almost ready for armament. Secretary of War William L. Marcy, however, avowed that the system was far from complete and asked for larger appropriations. Congress granted $1.3 million in 1846, but the Corps of Engineers reduced its request for the next year to one-third that amount, owing to the Mexican War.[17]

Congress appropriated "a much smaller sum" than the half-million dollars Totten asked for in 1847, so not much was accomplished. Secretary Marcy counseled against neglecting the subject, while Totten warned that it still would take years to complete the program. The new territories of the West then captured the War Department's attention. Budget requests for permanent forts were modest, and appropriations even more so. Meanwhile, the department created a board of army and navy officers to examine fortifications needs on the Pacific Coast. Totten fairly begged for money to resume work in 1849.[18]

There were separate Boards of Engineers for Fortifications for the Atlantic and the Pacific coasts by 1851. The eastern body warned that advances in naval steam technology made possible enemy fleets unlike anything seen before. That hyperbole gained the program little, as Congress appropriated nothing that year. Secretary of War Charles M. Conrad became more interested in coastal forts in 1852, when he asked for renewed appropriations. Chief Engineer Totten not only wanted to complete the system and extend it to the Pacific Coast but proposed additional work on the Atlantic and the Gulf.[19]

Fort appropriations resumed in 1853. That same year, Jefferson Davis became secretary of war, and he promoted the defense program. Steam navies and big guns, he said, made big forts even more valuable than before, while the growing wealth of the country increased the need for defense. Totten now had an ally, and over the next few years he started several new projects. True to his origins, from 1853 on he maintained that all recent technical and political developments justified the plans adopted four decades earlier.[20]

By that time, however, only sixteen engineers were available to manage work on over forty forts, the rest being busy with civil works. Moreover, only eleven of fifty permanent stations in the East were garrisoned, with the bulk of the army in the West or in Florida. In 1854 the Crimean War came to Totten's aid by proving that nations like Britain could invade a distant shore in force. Over the next eight years the government spent $14.9 million on fortifications—more than a third of all funds invested in coast defense between 1812 and 1861. Totten also continued to plan additional works.[21]

Construction moved along faster in the late 1850s than it ever had before. Conditions changed even faster, however. Major Richard Delafield, Corps of Engineers, returned from duty as a military observer in the Crimea in 1856 and declared that the demonstrated British ability to send large artillery expeditions overseas must lead to a redefinition of American defense. Secretary of War John B. Floyd thereupon called for moderation of the Third System. Large appropriations were not necessary, he said in 1859, because earthworks would suffice. He wanted to improve the defense of the whole coast against enemy landings, particularly where forts were absent. He told the engineers to study that problem for him.[22]

Lieutenant James St. C. Morton conducted the study, focusing on New York. The country in recent years had spent over $30 million, or $2 million a year, he pointed out, and still felt vulnerable. He recommended that the government (1) strengthen unfinished forts with earthwork batteries and "stop short" on construction of masonry forts of all kinds except those in very remote locations; (2) guard lines of approach to all ports by systems of entrenchments, where militia could rally against invasion; and (3) experiment with torpedoes (as fixed submarine mines were then called) as a means of channel defense. All that, Morton maintained, could be done nationwide at about one year's cost of current fort building.[23]

Delafield's and Morton's reports appeared in time to have some influence during the Civil War but not to alter the course of the Third System. Acting Chief Engineer René Edward DeRussy showed every intention of completing the system in 1859 and the next year called appropriations to that end "entirely inadequate." By that time the United States had one of the larger coast defense establishments in the world, although some forts were deteriorated and many were not armed. Masonry forts and heavy guns guarded all important harbors and invasion routes and were the keystone of the harbor defense system.[24]

More amazing was that the seacoasts were defended by a policy over four decades old. The Third System was never importantly revised. It was one of the longest-running public works ever undertaken in this country. That it also was the most unbending was owing to Chief Engineer Totten, who remained true to the vision of Simon Bernard. Together they left their mark on the coastal landscape. Only time would tell whether their imprint was a sign of strength or the dead hand of the past.

The Third System at Hampton Roads
Engineer Rivardi decided in 1794 that Hampton Roads should be covered by works at Old Point Comfort. Colonel Armistead had cause in 1813 to wish that that advice had been heeded,

when the British assaulted Craney Island. The Board of Engineers apparently determined from the outset of its planning that the place to fortify Hampton Roads was Old Point Comfort. On 21 April 1817 Armistead was ordered there to collect materials. On 25 July 1818 the Corps of Engineers let a contract for stone.[25]

The War Department wanted to work as much as possible by contract, and in 1818 Chief Engineer Swift entered five such arrangements. One went to Elijah Mix of New York, to deliver stone from quarries on the York River to Old Point Comfort and Rip Rap Shoals at Hampton Roads. The contract was straightforward enough. The price seemed low at the time and might have bankrupted Mix if the panic of 1819 had not changed the value of the dollar; the contractor made out well enough. Chief Engineer Alexander Macomb said in 1822 that "the contract to this period has been executed faithfully, and to the satisfaction of the department; and that the quantity delivered has, in every instance, been equal to the portion of the appropriation allotted to the contract."[26]

Christopher VanDeventer, chief clerk of the War Department and witness to the contract, was Elijah Mix's brother-in-law. He did not influence the contract, but matters did not look right when it became known that there had been no advertisement, merely invitations to bid for a few stone dealers. Worse, Mix was soon in financial trouble, and VanDeventer assumed part of his liability, with Mix assigning partial interest to his brother-in-law. VanDeventer, acting with Secretary Calhoun's knowledge but against his advice, later extended his share and by April 1819 owned half interest in the contract. Mix was in better straits by 1820, and VanDeventer sold his interest to other relatives, expecting a profit. The two then quarreled over who was entitled to payments from the War Department. Calhoun threatened to fire his chief clerk, until the relatives reached an accommodation in 1821, when the threat was withdrawn. Rumors had gotten out by that time, and a committee of the House of Representatives assembled to look into the affair. The legislators branded the Corps of Engineers' failure to advertise a "*singular* neglect of duty in the officers of Government."

Swift described the proceedings as "the allegation of the Radicals of Mr. Calhoun's alleged malversation in the (now become celebrated) Rip Rap contract with Mr. Mix, accusing the minister and the chief engineer, Swift, of partaking." The committee recommended suspending Mix's contract and requiring advertisement in all such actions in the future, but the full House refused to accept its report. The scandal then faded, only to be revived in 1827, when Calhoun endured forty days of hearings by a Senate committee. He and the Corps of Engineers were exonerated again.[27]

The Mix affair was only a diversion in the fortification of Hampton Roads. The Board of Engineers for Fortifications no sooner started work than Swift and Lieutenant Frederick Lewis went to Norfolk 3 April 1817 and then to Old Point Comfort, to report on positions from Lynnhaven Bay to Craney Island by way of Hampton Roads, to guard the roadstead in Lynnhaven Bay. Nothing came of that expedition. On 16 August 1817 Calhoun appointed a board of commissioners to cooperate with another from the Navy Department to develop defensive plans for the entrance to Chesapeake Bay. Swift and Armistead were assigned to the board, and by October Swift was on his way to Washington to join in its work. The members arrived in Baltimore 18 December, then went to Annapolis and on to Norfolk, which they reached 29 December. They surveyed Old Point Comfort the next day.[28]

The War Department had already decided in 1817 to build two works at Hampton Roads. One would be at Old Point Comfort and the other on the shoal called the Rip Raps, opposite Old Point on the southern side of the channel. Calhoun appointed James Maurice of Norfolk to supervise construction at what was now called Fort Monroe, on Old Point, and a year later to take charge of work at the Rip Raps, both until October 1821. Swift's army-navy commission sent the secretary of war a report prepared by Armistead on the defense of Hampton Roads. By 12 January 1818, according to Swift, the Board of Engineers had agreed to fortify Old Point Comfort and Rip Rap Shoals, and on 26 January the board assembled in Norfolk to confirm its findings. Swift sent Calhoun a preliminary report on proposed forts on the lower Chesapeake Bay, estimating that works at Old Point and the Rip Raps should cost about $3 million.[29]

Work seemed ready to begin. Swift ordered Lieutenant Colonel William McRee to Old Point Comfort in May 1818. The chief engineer followed him there with Simon Bernard and a navy officer, intending to resume surveys at Old Point Comfort, Gosport, the York River, and other locations. They played host on 5 June to President James Monroe and the secretaries of war and navy, who wanted to see the positions that had been selected for construction. Finally, on 1 August 1818 Secretary Calhoun refined the status of James Maurice, appointing him "Agent of Fortifications for Norfolk, Hampton Roads, and the lower part of Chesapeake Bay." Swift would handle contracting and disbursements, while Maurice superintended construction. A lieutenant soon arrived at Old Point Comfort to oversee construction of a wharf and anchorage.[30]

Swift reported in October 1818 that not all the fortifications of Chesapeake Bay had been determined, but that the selection of Old Point Comfort and Rip Rap Shoals was confirmed. The two forts would mount 250 cannons each and would cost $3 million together. He asked for $330,000 for contracts during 1819, to lay a foundation on the Rip Raps and another at Old Point Comfort. Work at the Rip Raps, he said, "is progressing," while "about two millions of bricks, and about twenty thousand perch of building stone, are collected at Old Point Comfort, under the care of an officer of engineers. The work at this position will commence next spring."[31]

A contractor began work at Old Point in March 1819, under the superintendence now of Major Charles Gratiot. By 1820 continued refinement of the plans had altered the project. Fort Monroe would mount 380 guns, with a rampart perimeter of 2,304 yards; Fort Calhoun, on the Rip Raps, would boast 216 guns in a perimeter of 381 yards. They were first described for Congress by Armistead:

> Fort Monroe . . . is a regular work, with seven fronts. Wharves, roads, machinery, workshops, and barracks have been built, and large quantities of materials collected, preparatory to the commencement of the work. All expenditures, except contingent, are provided for by special contracts.
>
> Fort Calhoun, on Rip Rap Shoal . . . is a tower battery, with three tiers of casemates, to be built upon a foundation, *a pierre perdue* [on broken stone], in a depth varying between one and a half and three fathoms. Forty or fifty thousand perches of stone have been applied to the formation of the

foundation, which now shows between two and three thousand perches above high-water mark. All expenditures, except contingent, are provided for by special contract.[32]

The War Department listed Fort Monroe as three-quarters completed in 1821 and the fort on the Rip Rap as one-third finished, with nearly $1 million spent on them. Meanwhile, the Virginia General Assembly authorized a transfer of land to the federal government in March 1821. Involved were 250 acres at Old Point Comfort and 15 acres at the Rip Raps. The deed, executed in 1838, reverted the land to the state if the United States abandoned it or used it for any purpose but national defense.[33]

Although construction was proceeding apace, the Board of Engineers did not present its first full report on the Third System until 1821. It therein clarified its objectives at Hampton Roads: "In the Chesapeake, the projected works at the entrance of Hampton roads have for object to close this road against an enemy, and to secure it to the United States; to secure the interior navigation between the Chesapeake and the more southern States; to make sure of a naval place of arms, where the navy of the United States may protect the Chesapeake and the coasting trade; to cover the public docks, &c. at Norfolk, and those which may be established in James river; and to prevent an enemy from making a permanent establishment at Norfolk."

The board observed that an enemy could land in Lynnhaven Bay and march inland. If his fleet was denied use of Hampton Roads, however, he could anchor only in Lynnhaven Bay, and his land forces could be turned and prevented from establishing themselves in Virginia or North Carolina. "The expense at which these results will be obtained," said the board, "is one million eight hundred thousand dollars—a trifling sum, if compared with the magnitude of the advantages which will be procured and the evils which will be averted."[34]

The board listed all proposed works according to priority and class (size). The two at Hampton Roads were among only six enjoying first priority, among a total of eighteen first-class fortifications. There were no second- or third-class works proposed for the area and none of second or third priority on the schedule of construction. What, then, of the remains of the so-called Second System and of construction during the War of 1812? They had not disappeared. The War Department spent significant sums to maintain the works at Norfolk between 1816 and 1824 and purchased Craney Island in 1817. In 1818 Fort Nelson hosted a garrison of eighty-eight officers and men, while Fort Norfolk had fifty. Fort Nelson still mounted thirty-seven cannons, Fort Norfolk thirty, and Crancy Island and Fort Powhatan had twenty and thirteen guns respectively.[35]

Forts Nelson and Norfolk declined after 1818, but the two places were not yet forgotten. "The existing forts, viz: Fort Nelson and Fort Norfolk," the Board of Engineers said in 1826, "serve for the defence of Norfolk and the navy yard. They are small and inefficient works, but may be useful as accessories to general defensive operations." The board had no interest in Craney Island or Fort Powhatan, while the War Department lumped Forts Nelson and Norfolk together as "Norfolk harbor, Norfolk, Virginia." Troops began to leave there for Old Point in 1823, and by early 1824 the two forts were abandoned. The War Department remembered them and Craney Island in 1826, when it listed them as "To be preserved, but not as part of the system." Regarding the condition of all three, however, "no details can be furnished." Fort Nelson vanished first; the navy took it over, and it was cleared away in 1827 when construction began on the Portsmouth Naval Hospital.[36]

A decade later, Chief Engineer Totten refused to count Fort Norfolk out of the plan entirely. "The existing fort, viz: Fort Norfolk," he said, "will aid in the defence of the city of Norfolk and of the navy yard." He felt the same way in 1848, when the Bureau of Yards and Docks asked that Fort Norfolk be transferred to the navy, which wanted to build a powder magazine there. Totten conceded that the fort was too dilapidated for further War Department use but still thought it might have some military value. He went along with transfer when the War Department imposed the condition that the fort not be demolished. On 14 September 1849 Fort Norfolk became navy property; it remained so until 1942, when it reverted to the Corps of Engineers, becoming headquarters of the corps's Norfolk District.[37]

That left only Craney Island still in War Department hands. The army let Armistead's fortifications wither away, but the Board of Engineers showed continuing interest in the place, despite dismissing it as "not part of the system." Beginning in 1822, the department added to the defenses of Hampton Roads works of the "Third Class, to be commenced at a remote period." They included rafts to obstruct the channel between Forts Monroe and Calhoun and new forts at Craney Island Flats, Newport News, and Naseway Shoal. The four items remained on the books until 1836, when they were demoted to a new "Fourth Class (Conditional)." Said Totten: "In the event of a great naval depot being fixed on James River, it might ultimately be desirable to provide additional strength, by adding works on the positions of Newport News, Naseway Shoals, and Craney Island flats."[38] Such talk was vain. There were limits to how much Congress would pay for fortifications, especially secondary and tertiary works. Fort Monroe and Fort Calhoun would be Hampton Roads' entire defense in the Third System.

"Great care has been manifested by the engineers in carrying on these works," Chief Engineer Macomb said of the projects at Hampton Roads in 1823, "and the execution of the workmanship is creditable to the superintending officer." Forts Monroe and Calhoun were indeed challenges, and not alone because of their complexity. The engineers were also pulled in many directions at once. There were four officers at Hampton Roads in 1826, when the work required five. The War Department planned to reduce them to three in 1828, two in 1829, one in 1830, and none by 1831, when it was hoped construction would be completed. The work was not completed then, for a variety of reasons. Appropriations were never consistent and were seldom timely. The engineers often had difficulties contracting for work and supplies and persistent trouble finding laborers. Contracting was temporarily abandoned in 1829 in favor of time-consuming open-market purchases of materials.[39] Finally, it seemed that every time Fort Monroe neared completion, the Board of Engineers added something to it, while Fort Calhoun refused to approach completion at all.

The Hampton Roads defenses involved some of the Corps of Engineers' leading lights before the Civil War. Joseph G. Swift, whose commission organized defensive locations in the area, and who was known as the "first graduate" of West Point, became chief engineer. Walker Armistead, who conceived the general plan of Hampton Roads' defense, succeeded Swift as chief. Simon Bernard, the chief fort builder, made Fort Monroe his special creation, while Joseph G. Totten, its ceaseless defender and world leader in fortifications design, also became chief engineer.

Hampton Roads continued to draw talent. In 1829 Lieutenant Andrew Talcott served as the chief assistant engineer at Forts Monroe and Calhoun. He was aided by lieutenants George Dutton and I. Mansfield. They all were under the direction of Charles Gratiot, "Chief of the Engineer department, inspector of the Military Academy, and superintending the construction of fortifications in Hampton Roads." Although Gratiot in 1830 relinquished disbursements to Talcott, he continued to hold responsibility for the projects there.[40]

Robert E. Lee, the future Confederate giant, served under Talcott in the early 1830s and often was in charge during Talcott's absences. His tenure was not without incident. Commanding General Macomb and Inspector General John E. Wool inspected Forts Monroe and Calhoun in 1834 and concluded that the former could get along without further attention from the engineers, then involved in furious squabbles with the resident artillerists over quarters and prerogatives. The War Department ordered the engineers to the Rip Raps, while the garrison would complete Fort Monroe. Talcott transferred elsewhere, leaving Lee in charge. Lee moved to the Washington office on 15 October and was replaced by Captain W. A. Eliason, who remained in charge when the engineers resumed responsibility for Fort Monroe's construction in 1835.[41]

Eliason held forth at Hampton Roads until his death in 1839, one of several officers who graduated first or second in their classes at the military academy and were assigned to Fort Monroe and Fort Calhoun. Those works remained a personal responsibility of the chief engineer (the only forts in the whole program that claimed that distinction), especially after Totten assumed office in 1838. The area was always shorthanded, however, with the engineers spread thinly around the country, mostly on civil works projects. By the 1850s the work was often in the hands of civilian project supervisors. Major John L. Smith assumed charge in 1856, but he was also supposed to direct work at Fort Washington, Maryland, and on rivers and harbors in Virginia.[42] Smith fell ill and was replaced in late November 1857, and died just over a year later. His replacement was Lieutenant Colonel René Edward De-

Russy, then acting chief engineer. His service in the crops extended back to graduation from West Point in 1812, and he was second only to Totten in the corps's seniority. He carried the Hampton Roads defenses into the Civil War.[43]

Building Fort Monroe

For Monroe, chief of the two works guarding Hampton Roads, was the supreme achievement of the Third System and of its designer, Simon Bernard. As designed, it was a regular work with seven fronts, covering sixty-three acres and surrounded by a wet ditch 8 feet deep and varying in width from 60 to 150 feet; it was supposed to house 380 guns. The interior crest of the main work ran 2,304 yards, and there were bastions at all salients. The main work concentrated the greatest firepower on the first, second, third, and fourth fronts, from casemate and barbette guns; the second and third fronts coordinated fire with Fort Calhoun. The landward fronts were armed on the barbette only, except for casemates in the bastions. There was also a casemated water battery outside the ditch on the fourth front. To oppose a siege or landward assault, the entire structure was further protected by a stout system of earthen and masonry outer works anchored by a strong redoubt.[44] Taken all together, Fort Monroe was a superb expression of military geometry rooted in the principles laid down by Vauban and his French successors—hence a "regular work."

Fort Monroe's construction was prosecuted for decades, according to the original plans but with continual refinements. Its situation was unique. Hotels began to sprout around it in 1820, and it became a popular resort. In 1824 Calhoun established the first military service school, the Artillery School of Practice, at Fort Monroe. Unlike most other big forts, it was garrisoned most of the time, with recurrent feuds between engineer and artillery officers. Moreover, the place was partly built by hired slaves or free blacks—laborers and craftsmen both—but they were in persistently short supply. Beginning in 1820, military convicts were assigned there, providing about 200 workers.[45]

Fort Monroe took form rapidly. Armistead reported in February 1821 that the fort was two-thirds completed. It did not look it, he said, because most work so far had gone into collecting materials. Masonry construction began in 1820, and one casemated work with a capacity of forty 32-pounder cannons was already complete.

He thought the entire job would be done in five years, with excavation of earth and masonry construction constituting most that needed to be done. Alexander Macomb said more in 1823: "Fortress Monroe begins to present a formidable appearance: the exterior wall, ten feet thick at its base, is carried on an average all round the place to the height of twelve feet, and a wet ditch surrounds the whole work. A battery on the covert way is constructed, capable of receiving forty-two pieces, and in the three fronts of the fortress on the sea side embrasures are partly constructed for eighty-four guns; so that, in case of necessity, a battery of one hundred and twenty-six heavy guns might readily be mounted for the protection of Hampton Roads."[46]

Fort Monroe ran far over budget by 1824, so estimates were raised from the original $816,814.95 to $1,159,522.17. The increase, said the War Department, was attributable to the addition of "contingencies not included in the original estimate, but found to be indispensable in the prosecution of the work." These included wharves, wells, cisterns, and temporary quarters. There was more, because a board of engineers had reviewed Fort Monroe's design. They started with fundamentals—stone and brick volumes had been erroneously calculated from the drawings, and cost estimates were off through arithmetical error. The engineers reduced the depth of the ditch to cut costs for excavation and haulage. Arches were reduced from 180 to 130 degrees, decreasing the volume of brick masonry. Full revetments replaced demirevetments on the scarps of two fronts, raising the costs of stonework. Finally, the original estimate for permanent quarters was too low, so the board raised it.[47]

"At Fort Monroe," Macomb said in 1824, "the progress of the operations during the year has been steady and satisfactory. The work is of great extent, and yet all parts of it have been, in a more or less degree, advanced; and in some parts the main walls have been completed. Additional permanent quarters have been built, and the construction of a permanent hospital has been commenced." That was none too soon, for the artillerists moved into the place that year. In October, Fort Monroe played host to the marquis de Lafayette, then on a tour of this country.[48]

Work continued in 1825, by which time Fort Monroe provided an education in construction management. "The operations are organized in a manner that admits of the various branches into which they are divided being conducted with the utmost regularity," Macomb reported, "whether conjointly with or independently of each other." Progress continued the following year, but in 1827 a contractor failed to deliver bricks. "This failure," said the chief engineer, "has occasioned a considerable loss of time, but the amount of work done during the past year is nevertheless very considerable."[49]

Fort Monroe was a lively place; when the engineer and artillery officers were not feuding over quarters, they socialized. They formed a theatrical society, which Gratiot treated to supper after every performance. He required the actors to wear their stage costumes during his festivities, which often were riotous.[50]

A limited supply of stone slowed masonry work in 1828, but much was accomplished nevertheless. Crews finished nearly all casemates on the water fronts, and a "considerable portion of the ramparts on the other fronts [were] formed." Gratiot predicted that his budget request for the next year would finish the job. Masonry work would have ended in 1829, but the death of the freestone supplier left some yet to be done. Gratiot's crews placed 60,000 cubic years of earth in the embankments of the outer works. However, he postponed putting the top part of the embankments above the terreplein and forming its parapet in order to give the masonry of the revetments time to consolidate. The terreplein, it had been decided, would now carry the top tier of guns, and pending its completion the work turned to the "outworks on the front of attack" or landward side. By 1831 the crews had formed the ramparts, excavated the ditch, and embanked the glacis. They also built a counterscarp wall to prevent erosion of the ditch, pointed the masonry, and fitted up casemates as quarters. Finally, materials were assembled to complete the water battery.[51]

Construction of the outer works included removing the rubble of eighteenth-century Fort George. Workmen unearthed an iron ring inlaid with silver, with a coat of arms showing a bear holding a globe or heart surmounted by a cross. The crest was "an eye with wings conjoined." No one knew what the relic meant, but it caused excitement in the neighborhood. The artifact may have borne a curse, because Fort Monroe soon was beset by problems. Crews finished the last 1,000 feet of counterscarp wall in 1832, pointed

the scarp walls on three fronts, and arched a "considerable portion" of casemated covert way. The exterior revetment was partly
constructed, nearly 1,000 feet of slope wall in the ditch was finished, and the ditch before three fronts was almost all excavated,
along with other outer works of a similar nature. Then in August
cholera hit the laborers, and the superintendent stopped operations. Gratiot complained about the interruption and groused
about the trouble the engineers had finding laborers. Nevertheless, he predicted that Fort Monroe would be completed in 1833.[52]

Fort Monroe received its name in 1832. It had never, apparently,
been given a title officially, although it was known usually as
Fortress Monroe from the beginning. That was changed to Fort
Monroe by order of the secretary of war. The work was by that time
under Talcott's supervision, with Lee actually in charge most of
the time. Lee feuded with the garrison officers, endured the cholera scare and labor shortages, and made slow going in 1833. An
offer of a 15-percent pay advance did not bring in enough workers
to finish the job. Nevertheless, "all the permanent parts of this
work were completed last year," Gratiot reported in 1834. All that
needed to be done was to install the gates and raise several parapets for which earth had already been piled: That year the War
Department sent the engineers to the Rip Raps, and the funds and
work were turned over to the garrison. "But for [that] circumstance . . . I should most likely have had the gratification of report-
ing it finished," Gratiot sniffed.[53]

The garrison—as it happened, greatly depleted by temporary
closing of the artillery school and transfers of units—did little
toward completing the work. The engineers next received appropriations for Fort Monroe in 1836 and 1837 but had no officer
available. Engineers inspected Fort Monroe in 1837, and their
findings bothered the secretary of war. He wanted the outer works
completed, but someone had leveled the mound of earth assembled
for them before operations ceased. As if that were not enough, the
garrison had loaded earth onto the parapets before the masonry
cured, and the walls had cracked in several places.[54]

The officers who inspected Fort Monroe in 1837 recommended
extensive repairs and completion of the work and observed also
that the bridge connecting it to the mainland was privately owned.
It ought to be purchased by the federal government, they sug-

gested. Meanwhile, Captain W. A. Eliason took charge at Fort Monroe and Fort Calhoun. He started preparations for the main construction work in 1838, beginning with a new wharf, contracts for materials, and the hiring of "a large number of laborers." The secretary of war promised that the work "will be prosecuted without interruption until the works are completed."[55]

The crews forged ahead. In one year they completed the stone masonry of front five; built forty-three concrete piers and arches to relieve the scarp on fronts four and five; carried up a breast-high wall 3 feet high along the curtain and flank of front four; installed eighteen relieving arches over the gutters of the casemates on front three, against the parade wall; inserted iron water pipes into the masonry in order to frost-proof them; took up, recut, and reset the coping of the scarp on front four; repointed some of the masonry and laid forty-two iron circles for gun traverses in the water battery, getting it "ready to receive its armament"; repaired the bridges over the ditch on two fronts; built a wharf in front of front two, and replaced some broken sandstone quoins with granite. While all that was going on, the government purchased the Mill Creek bridge, and Eliason discovered that it was thoroughly rotten and the road leading to it also needed repair.[56]

Operations then slowed to minor work on the fronts facing the water. Meanwhile, Secretary of War Poinsett feared that the fall of Castillo San Juan de Ulloa at Veracruz meant that stone fortifications could be destroyed by exploding artillery shells. He ordered experiments at Old Point Comfort, where the engineers built a stone wall and the artillerists fired shells at it point blank. Poinsett was pleased to learn that "the shells broke against it, making very little impression. No doubt, therefore, need be entertained of the ability of our building materials to resist hollow shot."[57]

Work resumed in 1841, but it was mostly maintenance, repairs, or correction of design errors. The casemates, for instance, were uncovered and paved over "with a cement believed to be waterproof," then recovered with earth. More relieving arches were installed to stabilize the scarp wall, and some of the plumbing was reconstructed. The outer works continued to take form, the breast-height wall rose further, and the Mill Creek road and bridge were rebuilt. The addition of three companies of troops to the garrison interfered with the engineers' activities, but worse happened in

July 1842 when the Treasury ran out of money and all army construction stopped. By that time guns had been mounted in front four, and four others were almost ready to receive their armament. The work remaining to be done was minor, said the chief engineer.[58]

Work revived in November 1842, after an August storm caused a "dangerous encroachment upon the beach opposite to one of the bastions." Besides the "contracted scale" of repair work and laying of pavement in gun rooms, the engineers did Fort Monroe's first beach erosion work in 1843 and built quarters for laborers. The miscellany of repairs and completion tasks continued the next year at the same reduced level, then increased in 1845, on grading and finishing of earthworks, paving, breast-height wall construction, shot furnaces, and masonry repairs.

So it continued, with the engineers and supervisors working crews everywhere, but always facing more yet to be done or older work to be redone or repaired. An especially futile task began in 1846 when a well was driven from 68 feet deep down to 225 feet without hitting good water. The well was down to 232 feet in 1847 when it met resistance too much for its 8-inch pipe. A seawall was completed that year, along with the usual earthwork and maintenance. Work continued into 1848, then halted for want of appropriations. It resumed in 1849, with a lot of new stonework as well as the usual repairs. The engineer in charge thought that about $40,000 would finish the whole project in one year, but he was allowed to request only half that amount. Building and rebuilding and repairing continued at a moderate level in 1850, and preparations were made to lower smaller pipe into the 8-inch well. Five-inch pipe was down to 283 feet within a year, but still without water, while work continued on completing new outer works and correcting ventilation problems in the magazines. By 1852 the War Department, having added extra touches to be applied, estimated that $75,000 would complete Fort Monroe.[59]

The engineers had their $75,000 in 1855 and tried to finish all structural incidentals. Quite a lot was completed in 1856, but the roof drains of many of the casemates clogged up and were removed, redesigned, and reinstalled with wider piping. Meanwhile, Major Smith reported that completion of the well "will require a further appropriation, as will also the prosecution of work on the redoubt, and making certain indispensable repairs of the main fort." There was also, according to Smith, a desperate need to rebuild the Mill Creek bridge again.[60]

Congress appropriated $10,000 for the well in 1857, but no offers were received to drill it by contract. Meanwhile, DeRussy replaced Smith and announced that Fort Monroe was almost, but not quite, done. According to Chief Engineer Totten, "This work is ready for the entire armament, as originally determined; but all the platforms for guns of heavier caliber designed to replace a part of those are yet to be provided, except the two built this year." The low level of construction, reconstruction, and repairs continued into 1861, when Fort Monroe, as ever, was almost but not quite complete.[61]

America's largest fort bristled with guns when the Civil War broke out. The engineers had done a magnificent job designing and building such a place in an uncongenial political climate. They had not done it within their original estimate, which they exceeded 100 percent. Nor were they ever satisfied that it was ready for attack. Fort Monroe was, nevertheless, impressive. It was also a decided contrast to its opposite number on the Rip Raps.

Trying to Build Fort Calhoun

Rip Rap Shoal was bound to catch an engineer's eye, if only for its name. One of the simplest feats of engineering is to build a foundation or stabilize a bank with a jumble of stone, called "riprap." The shoal on the south side of Hampton Roads was aptly named, for it was a pile of rocks, over which one to three fathoms of water flowed at high tide. It is not now apparent who first decided to build a fort at the Rip Raps. Armistead is a likely candidate, as he was most familiar with the area. In any event, it was settled on by the Board of Engineers and supporting commissions by January 1818. Work started along with that at Fort Monroe, with the Mix contract providing stone.

Fort Calhoun, as it soon was called, had many stepfathers, because its final design was the product of repeated adjustments. The essential form was settled upon in 1818 or 1819. It was to be a tower battery of three tiers of casemates, set upon an artificial island, or "mole." Its ground plan was semielliptical, arching toward the channel; its isolation from shore eliminated the possibility of siege or infantry assault, so the structure lacked the bastions and elaborate outer defenses of Fort Monroe. The interior crest ran 381 yards, and the fort would mount 216 guns, later raised to 232. It would cross fire with Fort Monroe on the channel.[62]

About half the stone for the underwater foundation was on hand by February 1821, and Armistead predicted emplacement of the remainder during the coming year. Then, he advised the secretary, it should be allowed to settle a year or two before the superstructure was started. He avowed that the fort could be completed three years after the start of construction.[63] The Corps of Engineers originally estimated the cost of Fort Calhoun to be just over $900,000. It saw no reason to revise that figure after spending over half the amount by 1823, because Chief Engineer Macomb reported good progress, with the mole 6 feet above water. He was still optimistic the following year: "It had been contemplated to lay the foundation of the walls of Fort Calhoun during this season, and arrangements were made accordingly, but they could not be carried into effect for the want of an officer to superintend them. No disadvantage, however, is likely to arise from the delay; on the contrary, if the mole . . . has not thoroughly settled, which is possible, although not probable, there will be afforded additional time for it to acquire the requisite solidity."[64]

There was still no officer available in 1825. However, appropriations were "advantageously applied in the foundation of the mole, in collecting materials for the superstructure, in the erection of cranes, and the completion of the permanent wharf to facilitate the landing of materials, and in the erection of buildings for workshops and quarters, and in other auxiliary preparations necessary to be provided previously to commencing the superstructure." The engineers had built an island, and then a small town upon it; the fort seemed about to rise.[65]

The marquis de Lafayette visited the Rip Raps and declared it a pleasant place. Thereafter Fort Calhoun enjoyed a reputation as an exclusive resort; presidents Andrew Jackson, John Tyler, and Abraham Lincoln sojourned there. A grand delegation of dignitaries, led by Commanding General Jacob Brown, arrived on 17 September 1826 to combine pleasure with business. They laid a cornerstone, and the construction of Fort Calhoun was underway.[66]

Laying of the structural foundation started immediately, but at first work could go on only at low tide. The engineers made "considerable progress" on the foundations in 1827 and continued to enlarge the mole. In 1828 Gratiot boasted that the foundation "is now of such extent and firmness as to justify the construction, during the next year, of the first or lower tier of the castle. The whole of the foundations are laid to a height which will admit of the work being prosecuted at all stages of the tides. The materials used are of a durable quality, and the work executed is substantial."[67]

The fort rose almost as fast as Armistead had predicted. The walls of the first tier were mostly completed by 1831, and work was about to begin on the second level. Then the trouble started: the engineers discovered that the weight caused subsidence in the foundation. They stopped construction and limited operations in 1831 to receiving materials, which they distributed so as to equalize the settling. The fort continued to sink, so the supervisors planned to follow the same course in 1832 and "continue it until a weight of materials equal to that which the foundations must finally bear shall have been accumulated on them, after which the construction may with safety be completed."[68]

The fort sank six inches in 1831, and three more in 1832. Gratiot said that the lessened subsidence "indicate[d] clearly a tendency in the pile to assume a fixed position," but he wanted "20,000 tons of stone to compensate for the subsidence of the mole during the last two years." He asked for money to pile on 26,000 more tons during the next year. By 1833 the engineers had decided that their real problem was in the natural bottom beneath the mole, rather than in it or the foundation. They added 12,500 tons of stone to the mole and 11,800 tons of building stone on or near the walls to "compress the substratum." Still the place sank, helped by more stone in 1834—all that was required to build the fort—along with 3,465 cubic yards of sand dumped into the fort interior to raise the terreplein. Gratiot thought another "25,000 tons of breakwater stone" would solve the problem at last. That, he believed, would compensate for the subsidence. The plan then was to let the thing sit until an equilibrium established. The engineers added twice as much stone in 1834 as in 1833, while the annual subsidence was only one and a third times the 1833 figure. They hoped fervently that the irregularity of the settling was "fast disappearing." Meanwhile, their wharf blew away in a storm.[69]

Secretary of War Cass wanted to suspend all work on Fort

Calhoun in 1835. He relented when the engineers brought him evidence that the sinking would not further endanger the work. More stone was distributed over the foundations that year, giving an extra 20,000 tons beyond what the fort would weigh. Gratiot conceded that the place was still subsiding, but "in a continued decreasing ratio, and should there be no evidence of a contrary nature by next spring, it is proposed to resume the construction of the walls."[70]

Gratiot had to show some progress in order to save the project, so operations resumed at Fort Calhoun in July 1836. Workers removed the piled stone, carefully keeping the pressure equilized. The engineers hoped to resume masonry construction in March 1837. However, the relief of the loading caused the substrata to rebound unevenly, and an exasperated Gratiot ordered the weight reapplied, "and when this is done, it will be prudent to leave it there for some time longer before recommencing the works." His impatience had backfired.[71]

Reloading was interrupted repeatedly by labor shortages. Most of the work force at Hampton Roads had been slaves hired from their owners, and the engineers could find only twenty-five of them in March 1839, increased gradually to fifty-two by September. They piled scores of tons of stone in 1839 and 1840, massively increasing the loading of the walls and piers. "The increased subsidence since the recommencement of the provisional loading," said Totten, "shows conclusively the importance of this operation, and the necessity of continuing it until a load somewhat greater than the whole weight of the work shall have been placed upon the foundations."[72]

Totten persuaded the secretary of war in 1841 that delay of Fort Calhoun meant that Hampton Roads required a "strong field work" at Willoughby Point. The secretary asked for an appropriation but suspended its application "until required by circumstances." Meanwhile, the reloading was completed in 1841, and the foundations carried more weight than they would when the fort was finished. Crews worked until May 1842 leveling and adjusting, then withdrew to let the place rest. Subsidence declined to seven-eighths of an inch that year, but Totten decided to wait until it stopped altogether.[73]

The settling declined further in 1843 and ceased almost entirely in 1844. Totten asked for appropriations then, proposing to purchase granite and distribute it where it would be convenient to the stonecutters and add to the loading. However, no work other than studies of the subsidence went on the next year, when it was determined the timbers sustaining the loading were decaying. The mass of piled stone "is evidently pressing with great violence upon the piers," although Totten believed "at any rate, it is very clear that no actual construction can be commenced until that period, which can be reached only by this action of an adequate weight."[74]

Subsidence measured three-quarters of an inch in 1846, making Totten optimistic that it would soon cease. The continued rotting of the support timbers alarmed him, and he asked permission to use Fort Calhoun's unexpended funding to replace them. Nothing happened the next year, or the year after that. However, in 1848 subsidence was about a quarter of an inch, and the supervising engineer decided that it was almost at an end. He proposed to remove the loading by degrees, to begin construction as soon as appropriations were available. He was not allowed to ask for money for several years, however.

The place was still sinking in 1850, and Totten hesitated to remove the loading. He was more hopeful two years later, and in 1853 declared the subsidence "at an end," but he asked for no appropriations until 1854. Crews returned to the mole in 1855 and repaired and rebuilt their barracks. In June they began to take down the loading. No new movement of the foundation occurred, and work went on in 1856 and 1857. The last loading was removed in May 1857, and a contract was let for stone for the lower masonry courses, beginning in November. Gangs cleared away the remains of the original Fort Calhoun, finished the pier, built cranes, and began cutting and laying granite. By 1861 the first tier of casemates was essentially completed and work had begun on the second level. In the press of the national crisis the place was made ready to receive guns. New barracks and other facilities were erected, and Fort Calhoun was prepared to host a garrison. The war interrupted progress, and only a little stone was laid on the second tier.[75]

The revived Fort Calhoun featured something new in fortifications design and uniquely Totten's creation. He spent decades studying the problem of the minimum opening of casemate embrasures that would most protect guns and crews and least restrict movement of the guns. By the 1850s he had designed embrasure throats composed of multiple iron plates—the first use of metal armor in coast defense structures. They were installed on a number of Third System forts, including Fort Calhoun, and later acquired iron shutters for further protection. Totten's idea was adopted by the Russians and the British and spread around the world.[76]

The iron embrasure throats still are present on what remains of Fort Calhoun. Those on the scarcely started second tier stand forlornly as if on a widow's walk, looking out over Hampton Roads in memory of the stillborn castle on the Rip Raps. The work never saw completion. Fort Calhoun was the greatest frustration of the Third System.

War and the Future of the Third System

Early in 1861 Richard Delafield issued the final report of the commission of military observers he had accompanied to the Crimea. His text included studies of fortifications, fieldworks, and all other facets of war. Delafield and his colleagues had seen things that suggested the days of masonry forts were numbered.[77]

Secretary of War Simon Cameron feared that the secession crisis would make the United States a target for foreign powers. He expected the great forts to hold such enemies at bay and wanted them completed: "Aggressions are seldom made upon a nation ever ready to show to the world, that while engaged in quelling disturbances at home we are able to protect ourselves against attacks from abroad."[78]

The "disturbances" soon were out of hand. The unguarded forts in the South fell into rebel hands, except for Fort Pickens, Florida, and Forts Taylor and Jefferson in the Florida Keys. To the horror of the engineers, Fort Sumter, at Charleston, South Carolina, surrendered to hostile fire. Congress appropriated $1.4 million for forts in 1861, then $5.3 million in 1862 and again in 1863, along with substantial sums for temporary works. The Corps of Engineers was drained of officers serving in the field, but it pressed ahead on forts, with about a quarter of its officers working on permanent fortifications during the last full year of the war.[79]

Pursuing the Third System soon became an unjustifiable diversion of effort from the war. Worse, it looked futile. Thomas Rodman had overcome technical problems that limited the size of cannons. With his designs, smoothbore gunnery attained its highest level, the 15-inch Rodman, the most powerful cannon in the world. Almost simultaneously, former ordinance officer Robert P. Parrott achieved great strides in cannon rifling, followed by new ammunition that made muzzle-loading a rifle quick and simple. Explosive shells, never before a major type of ordnance, abounded in type and size. The new heavy weapons doomed masonry fortifications. The giant smoothbores appeared first to favor the forts, because they threatened enemy ships. Ships, however, grew iron-armor skins. Worse, big Rodmans could smash fort walls with iron balls of unprecedented weight.

The rifled gun was the real threat, however, and one which could be carried at effective caliber on shipboard as well as on land. Rifling, which twisted the projectile in flight, imparted greater range and accuracy, permitting attacking ships to stand farther away from their large targets, the forts. Rifled projectiles were elongated, placing the entire mass behind a small nose. They penetrated fortification masonry and exploded inside with a shower of brick and stone splinters. Union gunners retook Fort Pulaski, Georgia, in 1862 by blowing holes in its walls with rifled field guns, and similar demonstrations followed elsewhere. Army engineers reexamined old-fashioned earthen embankments and found them effective on the battlefield and at inland strongpoints.[80] The War Department appointed a board of engineers in 1864 to examine seacoast defenses. Its members were recalled to field duty before they could render a detailed plan, but they declared that earth for ramparts and parapets would be the most economical material to resist the new weaponry. The Corps of Engineers also wanted to prepare the coastal forts to mount the heaviest and most suitable gunnery for use against ironclads.[81]

Studies and construction continued in 1865, along with repairs to recaptured fortifications. Then the secretary of war halted construction until plans were offered to modify the southern forts to take the new armament. Delafield, now chief engineer, reported that his department had been building batteries for rifled artillery and giant smoothbores to guard against Confederate ironclad raiders. Three Corps of Engineers officers spent the last months of the war fortifying thirteen harbors to that end. Where before they had built gun platforms for 42-pounders or smaller, now they must prepare for 100-pounder rifles and 10- to 15-inch smoothbore guns. Looking to the postwar future, Delafield detailed several boards of engineers to consider what modifications would make all forts impervious to the "powerful armaments that European fleets, singly or combined, may be enabled to bring across the Atlantic." Accordingly, he wanted further high appropriations for coastal defense, sounding very much like Bernard, Gratiot, and Totten before him.[82]

Delafield was a brilliant man and one of the earliest interpreters of the Crimean War and the American Civil War. However, he had been in the Corps of Engineers since the inception of the Third System, and he shared the corps's attachment to its grand vision. He failed to appreciate the full nature of the changes occurring in military and naval science. Merely piling earth or attaching iron plates to the old forts would not sufficiently answer the new technologies.

The Third System had been conceived when navies comprised wooden sailing ships. The ships were durable structures, armed with equally stable weaponry. Smoothbore cannons generally fired solid iron shot; fighting ships threw iron balls at each other until something broke. Success in naval fighting depended upon good seamanship, allowing a captain to throw more iron at his opponent than the foe could return, as by sailing across his bow. Forts had all the advantages over ships in those days. They were more stable gun platforms, not at the mercy of the winds, so they could throw heavier balls of iron from larger guns than could be carried on the pitching deck of a ship, and do it more accurately. Furthermore, a wooden ship was likely to break under fire sooner than a fort.

All that changed by the end of the Civil War. To begin with, ships increasingly were powered by steam, giving them great freedom of movement—one of many developments ignored by the engineers during the course of the Third System, and even during the Civil War. The successful Union assaults on New Orleans and Mobile during the war demonstrated that with acceptable losses, a fleet could steam past fortifications to take the cities they guarded.

Moreover, after 1862 ships increasingly were clad in iron, often better protection than several feet of masonry. And the new gunnery gave ships weapons that could damage masonry walls, from increasing distances. The forts could adopt the same weaponry, with still heavier guns than ships could carry, but the balance was bound to be redressed. The Third System forts were neither located nor designed for battles that would move farther offshore with each new advance in gunnery.

Although its future was dubious in the 1860s, the Third System nevertheless had established an enduring legacy in American military defense—and in governmental programs generally. It was launched with the assurance that its designers knew exactly what they would do and what it would cost. The cost estimates proved far too low, however, because costs were impossible to predict over long periods with any accuracy and because unexpected conditions—as at Fort Calhoun—could disrupt all calculations. Moreover, the engineers pointedly and repeatedly claimed that maintenance (or rehabilitation due to lack of maintenance) was not a cost factor in the program—a result either of inexcusable ignorance on the part of structural experts or of deceptive bureaucratic budget seeking. Once the shortfall in estimates was apparent, the government was committed to the program and must continue it, even as costs rose further in response to refinements of design or addition of new forts.

Congress played its part in raising the expense of the Third System. It repeatedly displayed greater enthusiasm for the adoption of a program than for its prosecution and maintenance. Funding, over the life of the Third System, was erratic and fitful; the lateness of annual appropriations was a continual source of inefficiency. Moreover, Congress made the engineers victims of their own success in proposing a comprehensive, four-part system in the Board of Engineers program, for the most enduringly successful part of the Third System was its sanction of river-and-harbor improvements and other civil works at the hands of the Corps of Engineers. By the 1850s the civil works program was a great drain on the budget and manpower of the corps, persistently diluting its resources and distracting its concentration on fortifications.

The Third System established its own momentum in the War Department, carrying along each politically appointed secretary in turn. Lewis Cass was an exception, but even he was unable to sway the Third System from a course begun in 1817 toward adaptation to new technological and fiscal conditions, or even a more realistic assessment of the true military need. The Third System had a life of its own, thoroughly true to its self while the world changed around it.

And the world had changed by the 1860s, no matter how much the engineers wanted to deny the fact. The nature of the changes was demonstrated more than anywhere else at Hampton Roads, within sight of the Third System's grandest monuments. Forts Monroe and Calhoun and their type were the best the engineers could offer in 1817. Now they were not enough. Some new system of defense must be devised, amid the startling military innovations that appeared every year.

Confederate fieldworks around Petersburg in 1864. The siege of Yorktown in 1781 looked much like this. (National Archives negative 111-
B-373)

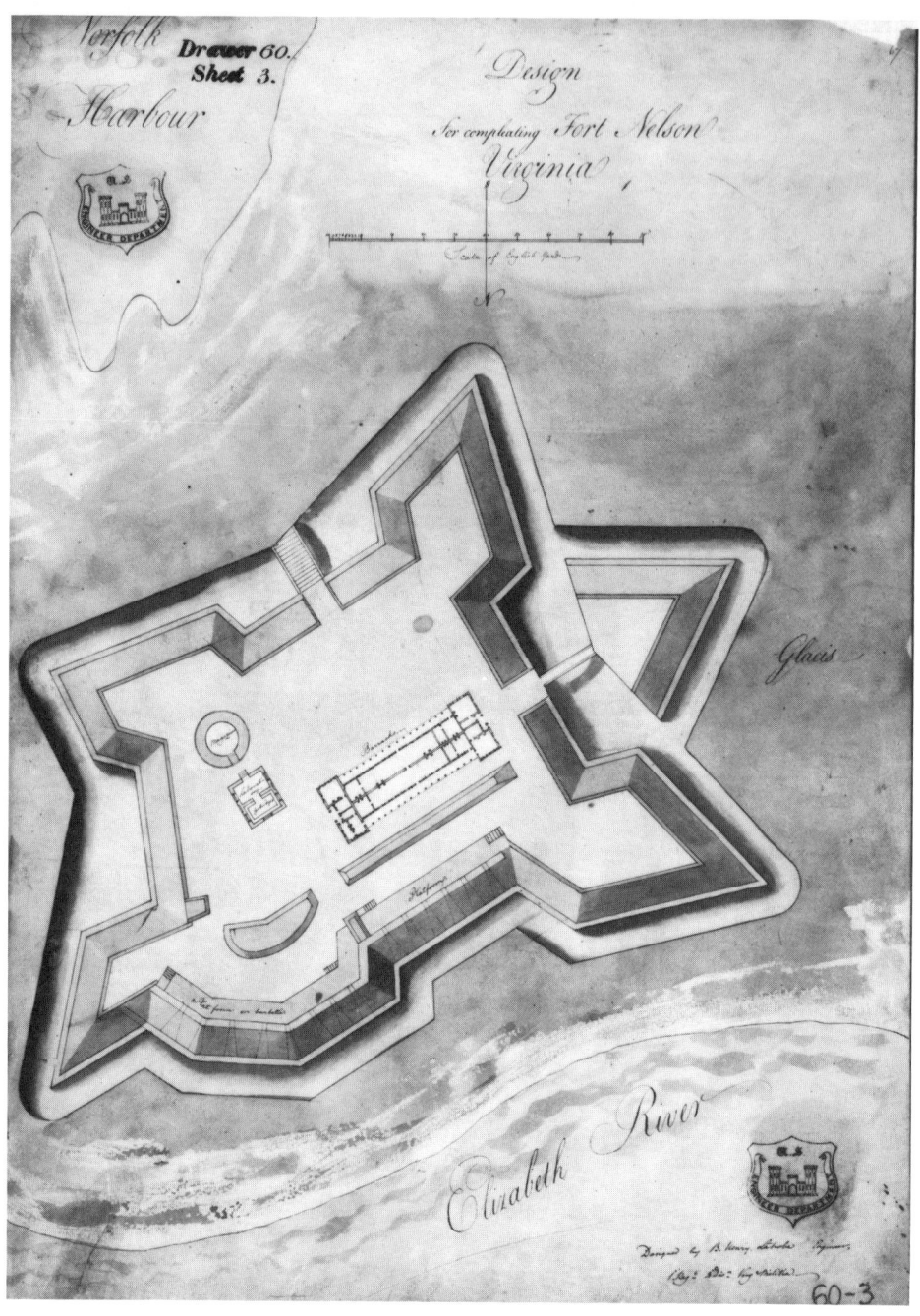

Benjamin Henry Latrobe's design for the reconstruction of Fort Nelson,
1798. (National Archives drawing Dr60Sh3)

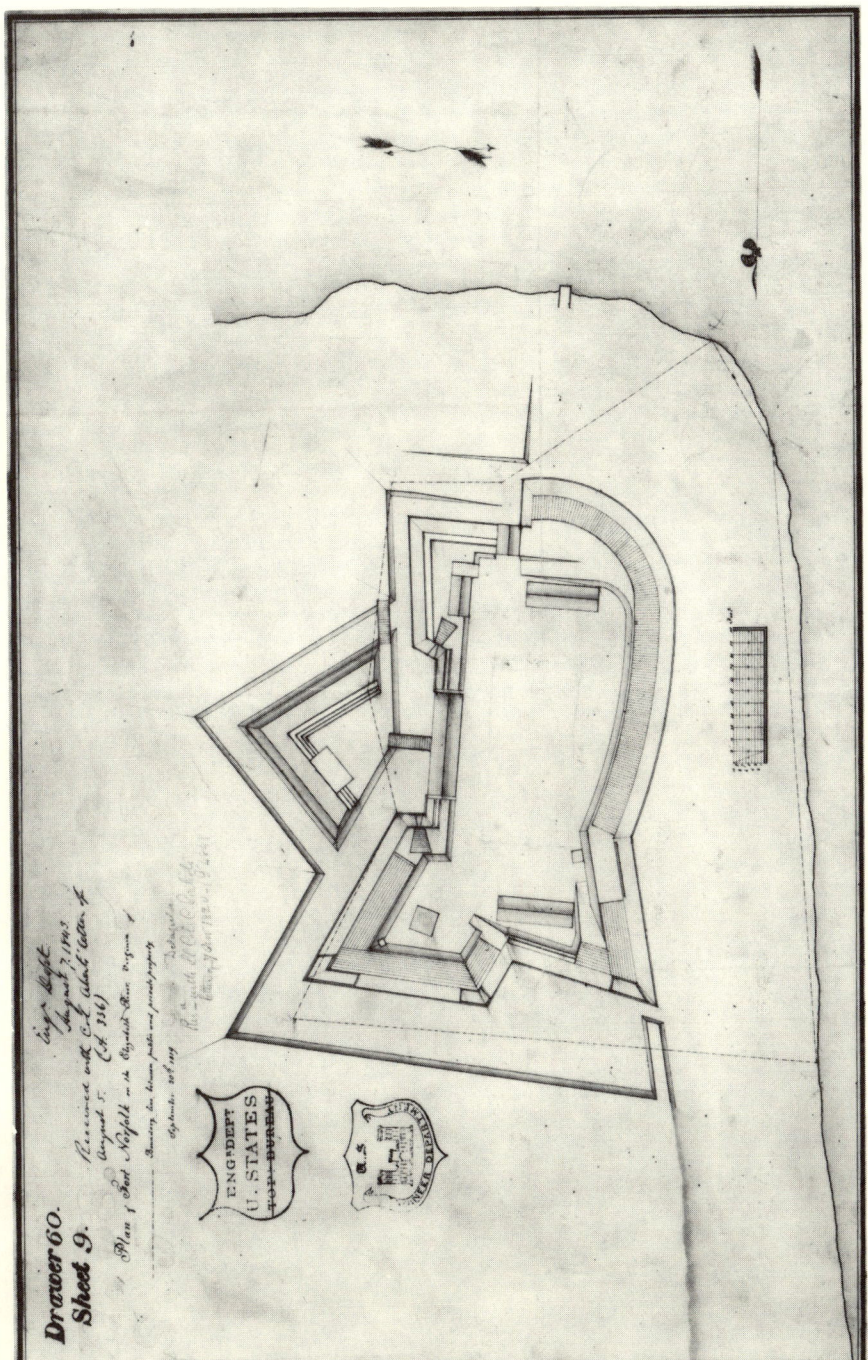

Fort Norfolk in 1819. The curved facade of the main battery is distinctive of the Second System. (National Archives drawing Dr60Sh9)

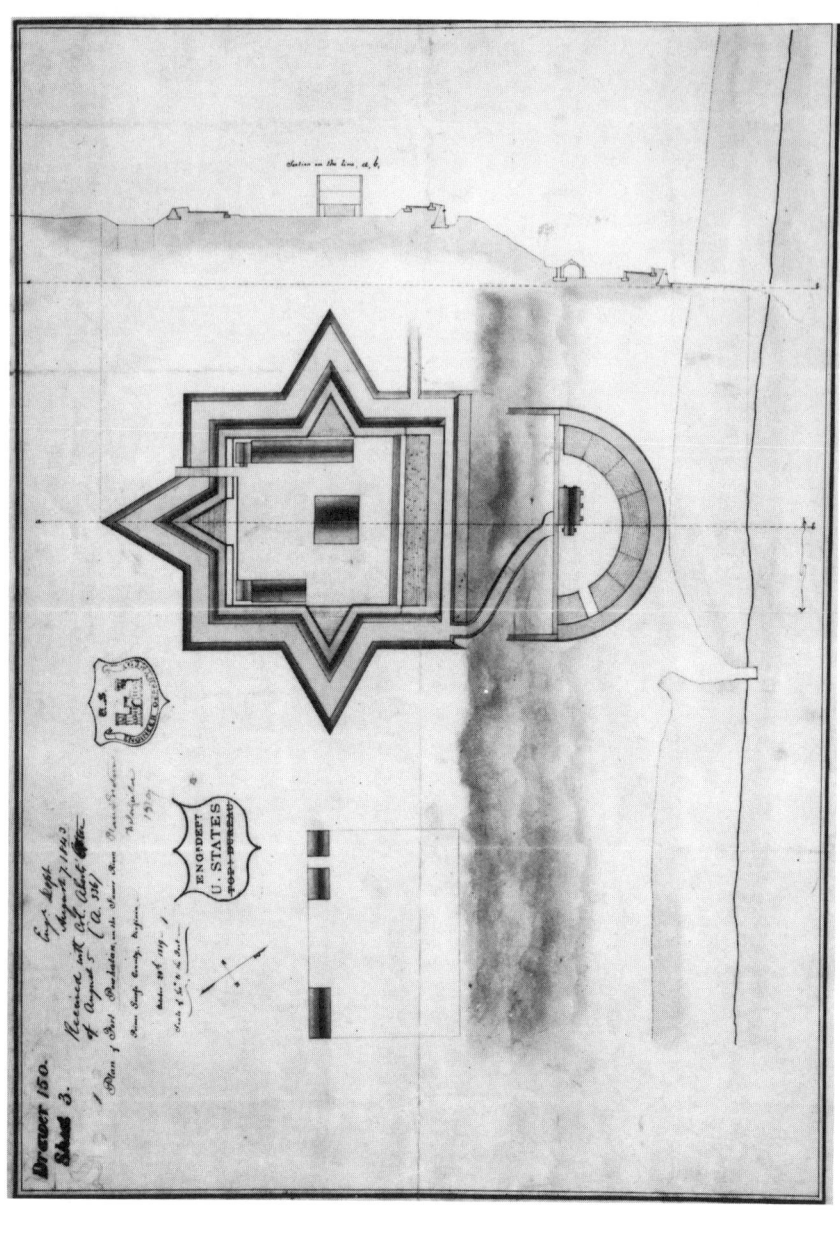

Fort Powhatan at Hood's, 1819. The water battery is below the bluff on the riverbank, protected by the fort above. (National Archives drawing DR150Sh3)

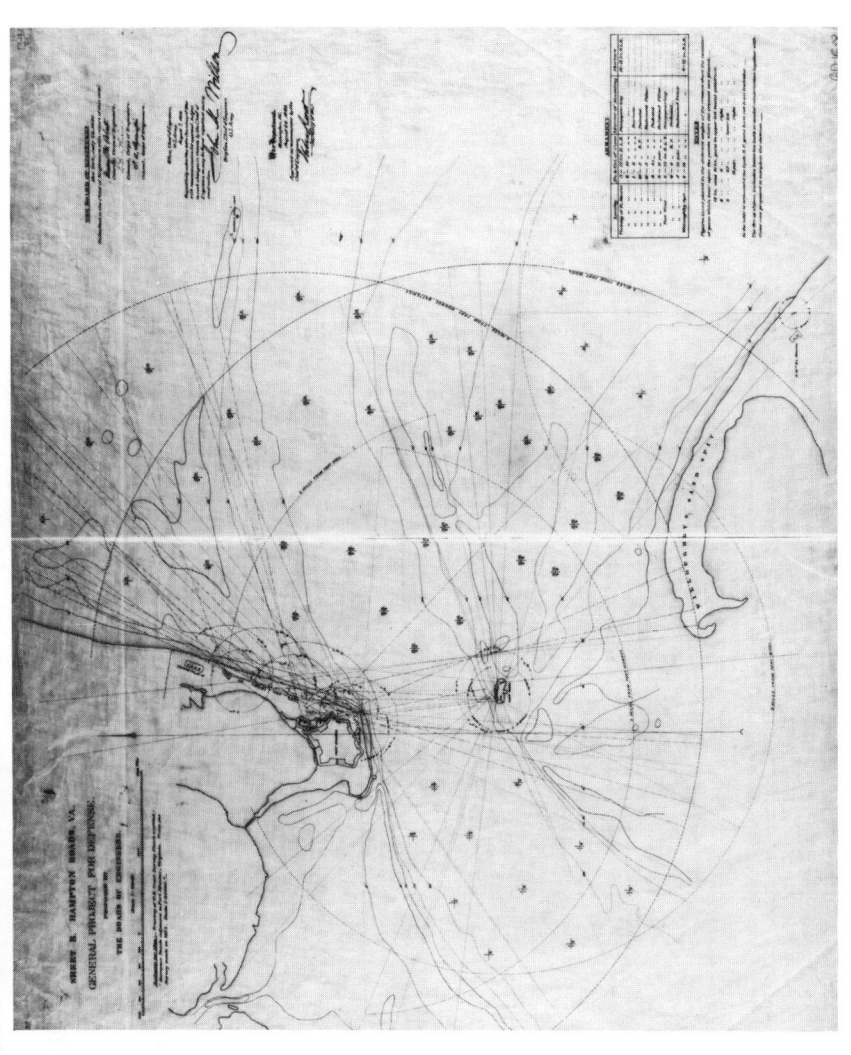

Fort Monroe, *upper left center*, Fort Wool (Calhoun), *left center*, and their relation to Hampton Roads, *left*, 1899. Hampton Roads' defense rested on these two locations for more than a century. (National Archives drawing Dr60Sh16-8)

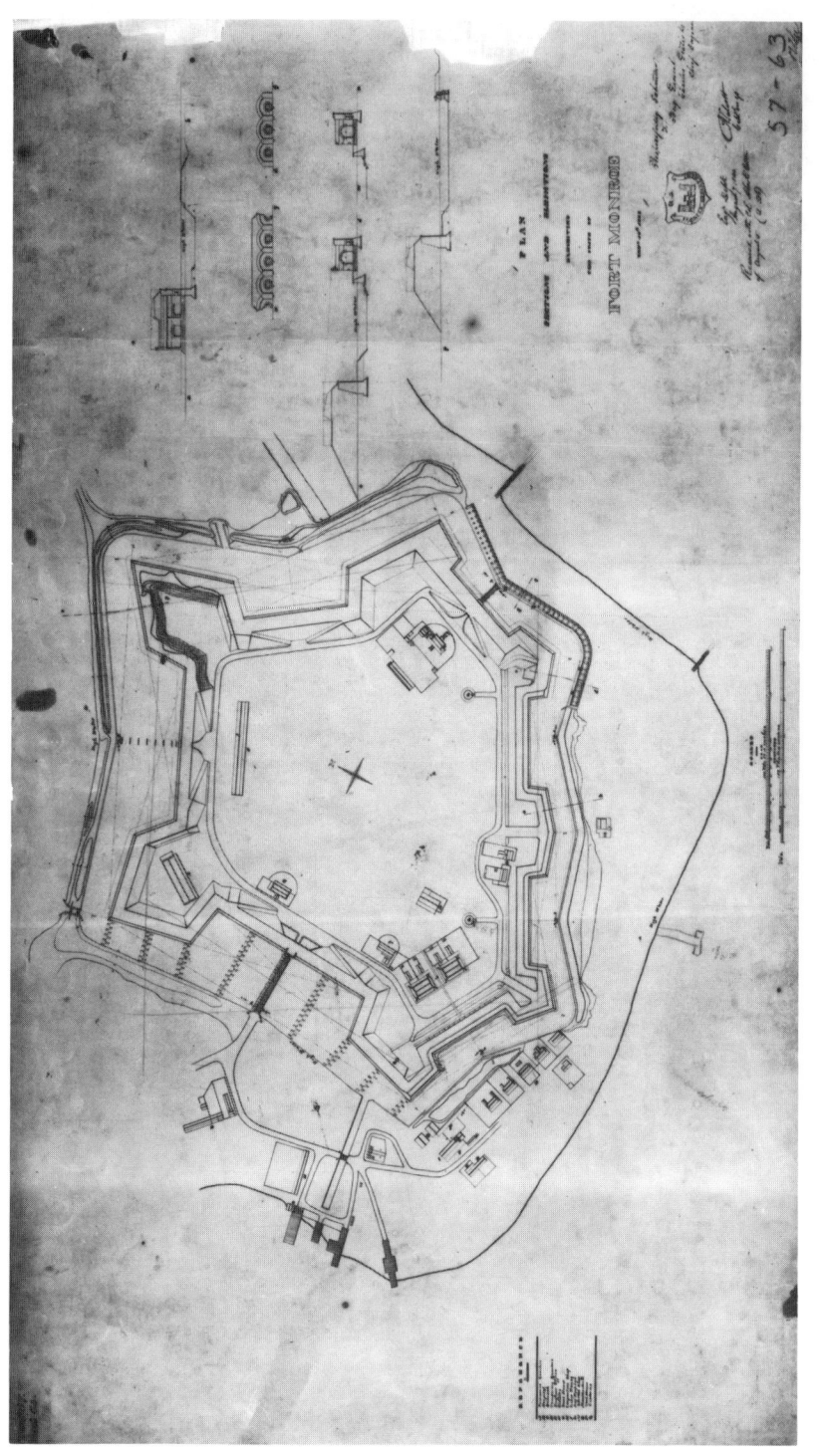

Robert E. Lee's depiction of the condition of Fort Monroe, 1832. Note the water battery outside and along the ditch southeast of the main fort. (National Archives drawing Dr57Sh63)

Fort Monroe, Old Point Comfort, and the Hygeia Hotel, in a drawing made just after the Civil War. The water battery is on the opposite side next to the flagstaff. (National Archives negative 165-C-7)

The Board of Engineers' plan for developing the foundation at Fort Calhoun, 1821. The engineers expected horizontal instability to be the major problem, but vertical instability repeatedly ruined their plans. (National Archives drawing Dr59Sh6)

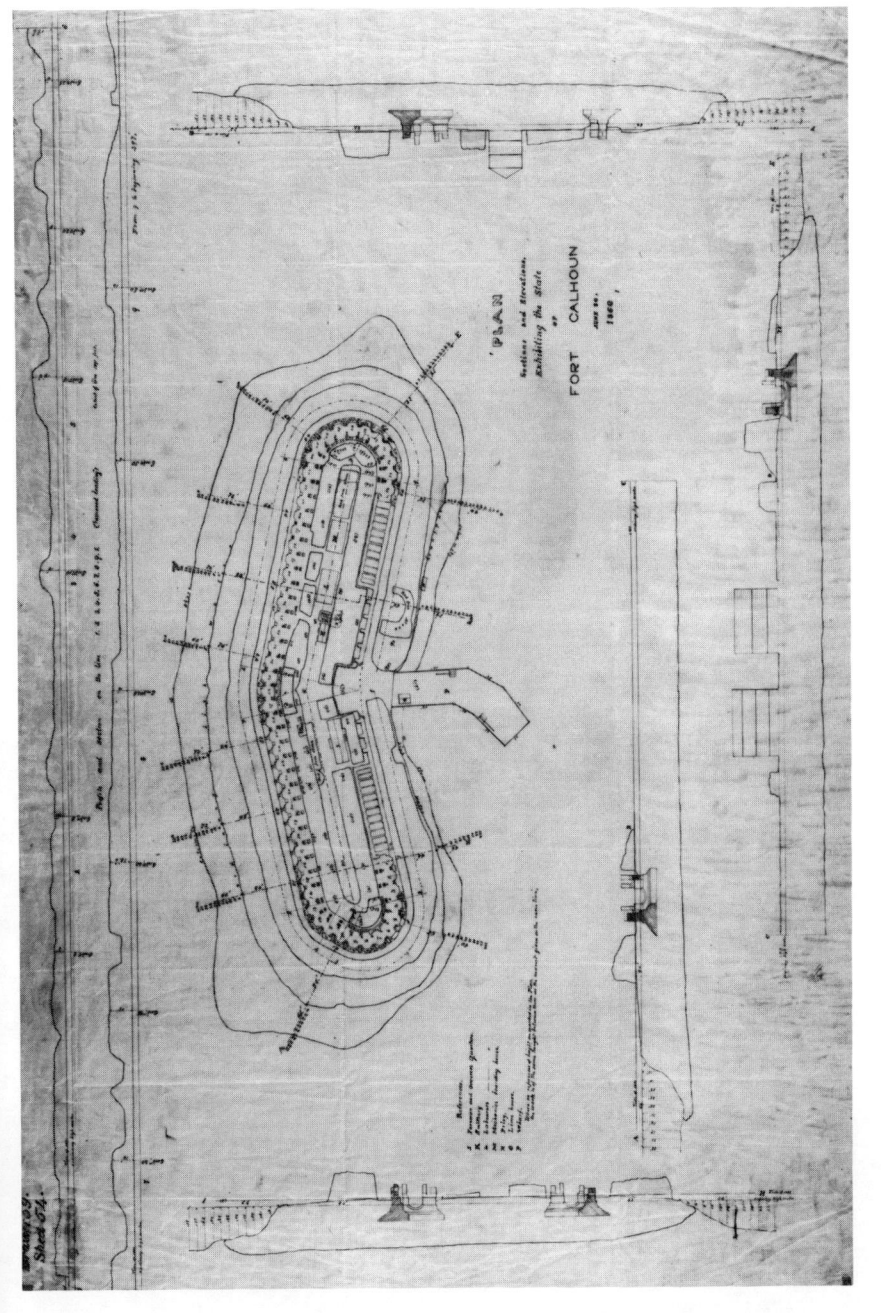

Fort Calhoun at its greatest height, 1860. The castle on the Rip Raps was only about a third of what the engineers had planned it to be. (National Archives drawing Dr59Sh54)

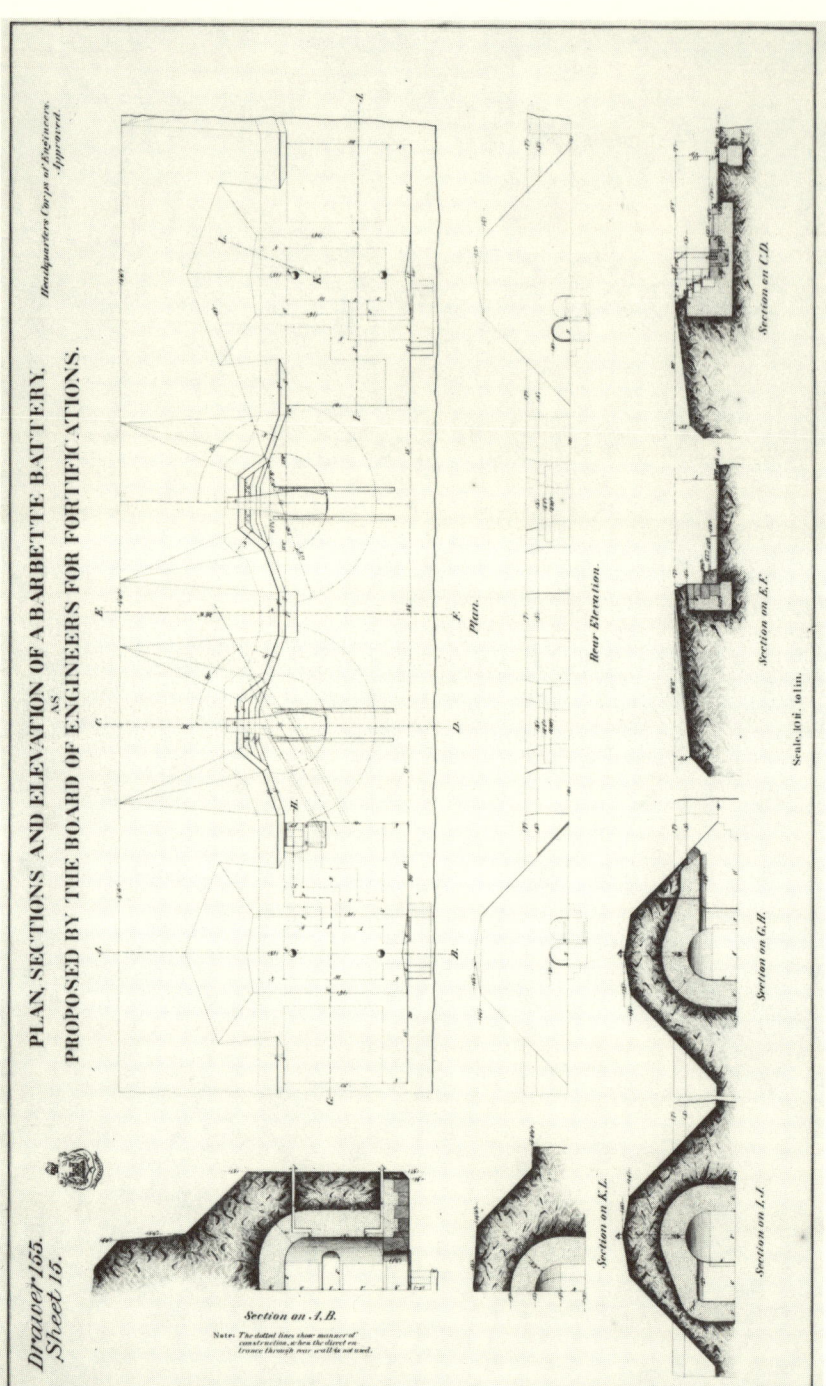

The Board of Engineers' design for barbette batteries in 1869. Because iron was too expensive, the engineers relied on massive earthen parapets and traverses. (National Archives drawing Dr155Sh15)

Rear side of the water battery at Fort Monroe, 1868. Note the iron shot stacked behind each casemate and the front-pintle barbette guns beyond, all obsolete. (National Archives negative 165-CN-2684)

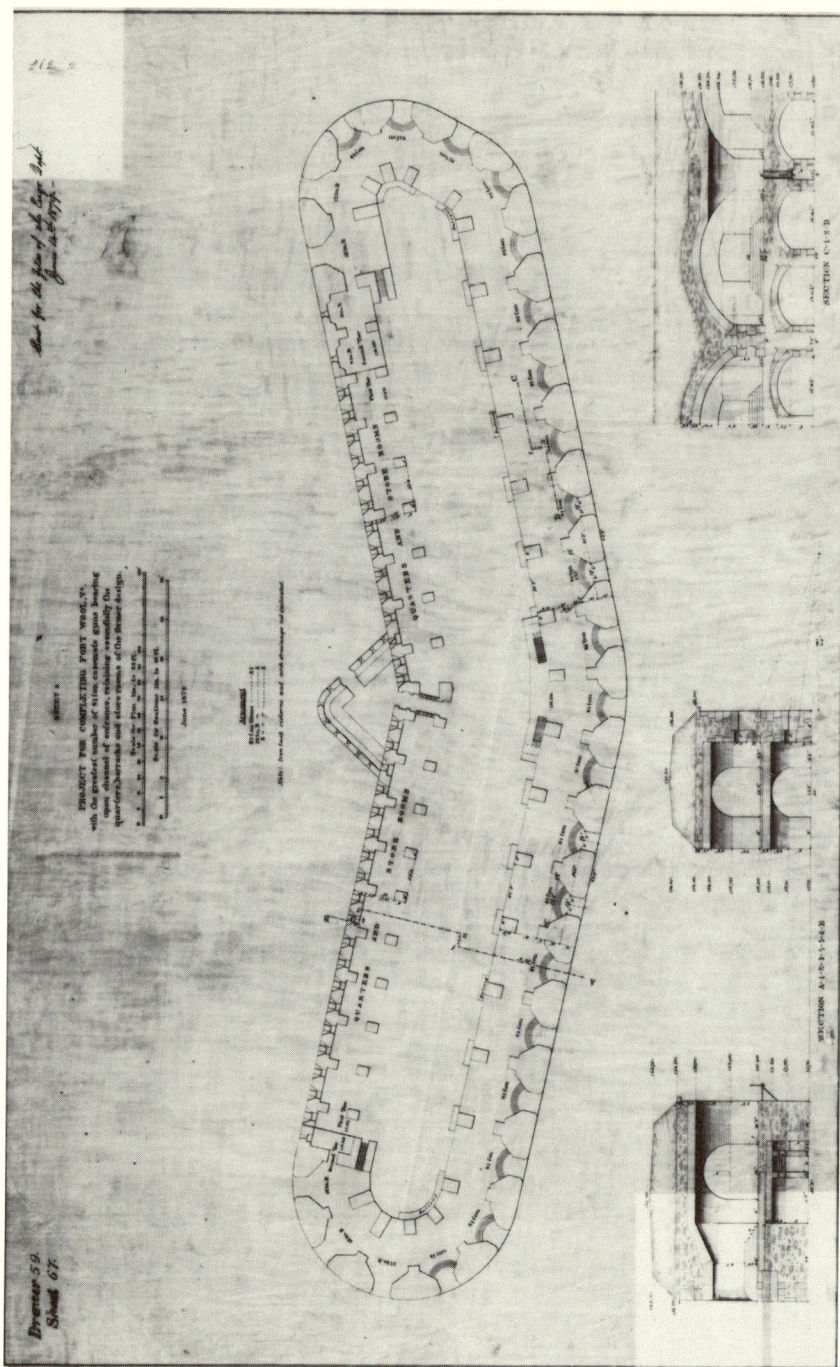

The 1879 plan for completing Fort Wool, "with the greatest number of 81 ton casemate guns bearing upon channel of entrance." This plan was rejected. (National Archives drawing Dr59Sh67)

Practice firing of a 12-inch battery on disappearing carriages at Fort
Monroe, 1918. The gun in the distance has just fired and is descending
to the loading position. The gun in the foreground is being reloaded.
(National Archives negative 111-SC-55954)

Firing the forward tubes of a 12-inch mortar battery, Fort Monroe,
1918. Note the projectile in flight and the men's reaction to the terrific
shock. (National Archives negative 111-SC-55950)

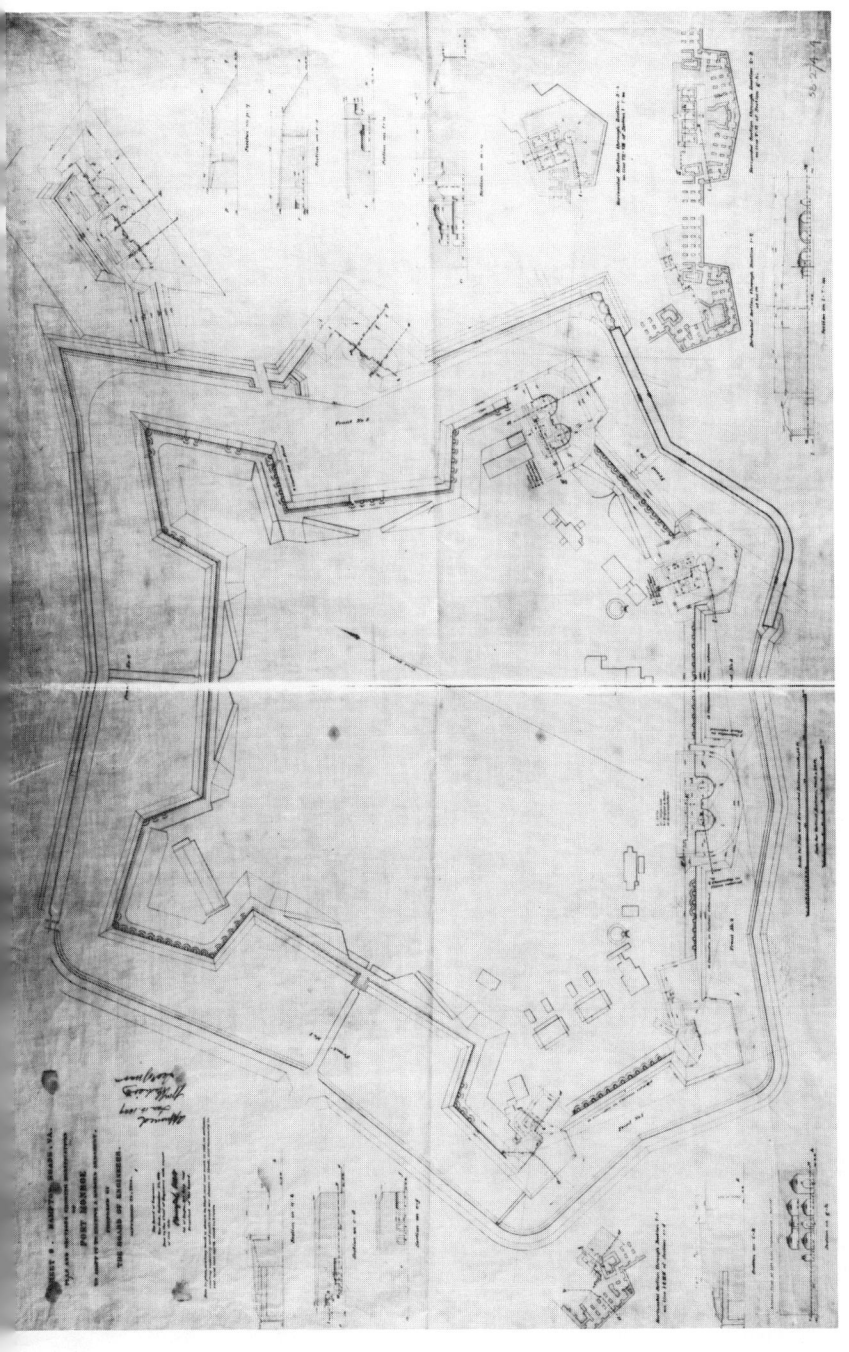

Plan for the modernization of Fort Monroe to accommodate advanced weapons, 1888. (National Archives drawing Dr58Sh274-4)

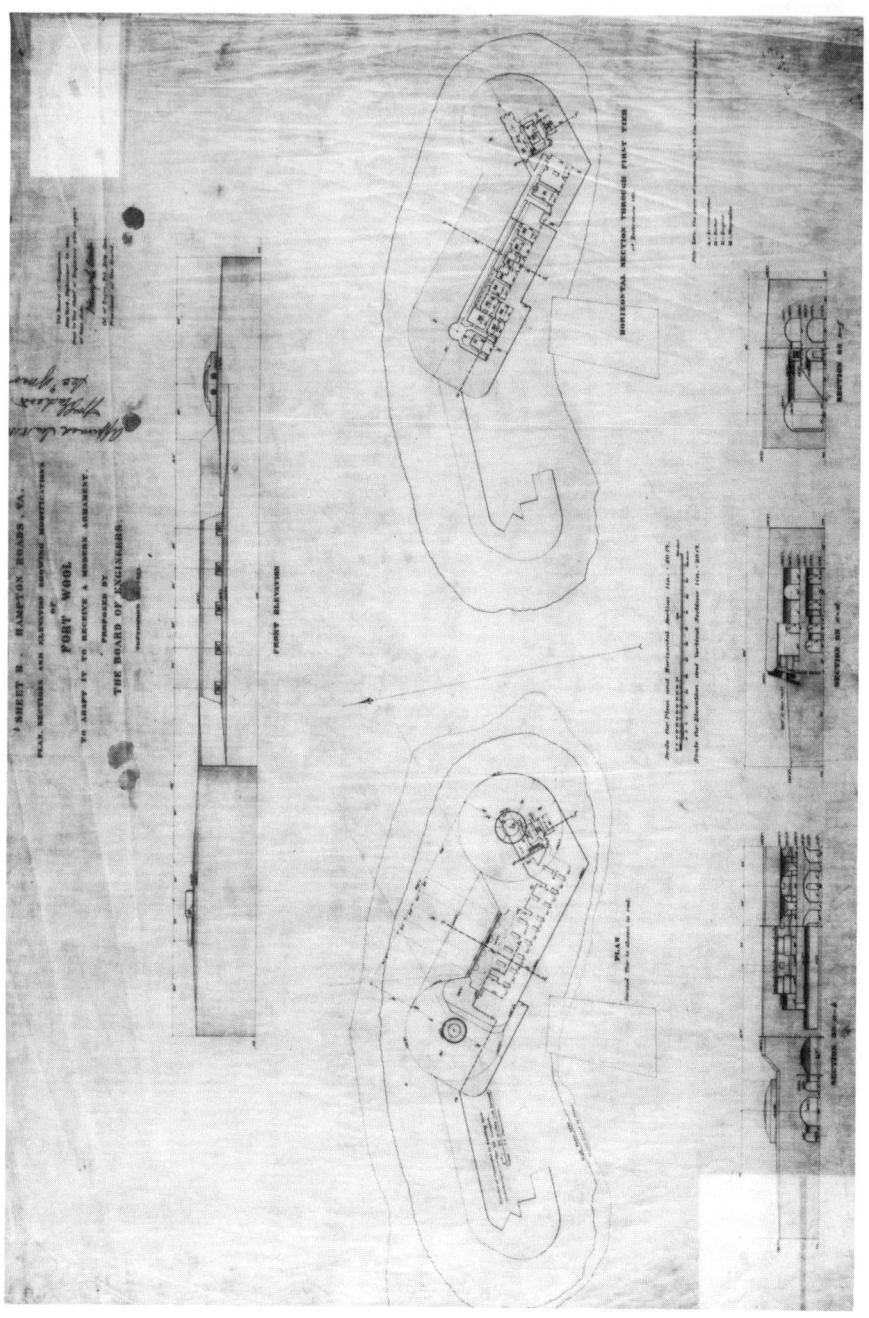

The first Endicott-period plan for Fort Wool, 1888; notice the two-gun revolving turret. (National Archives drawing Dr58Sh274-2)

Rear view of Endicott-period emplacement under construction at Fort Monroe in the 1890s. (National Archives negative 77-F-58-251-4)

Work in progress on the mortar batteries north of Fort Monroe, 1897. The wooden forms are being assembled before the placement of concrete. (National Archives negative 77-F-58-256-2)

Fort Monroe from a balloon, looking across Hampton Roads, early twentieth century. The Endicott-period emplacements are behind the beach
line at left, from Battery Church at the bottom to Batteries Parrott and
Irwin at the point. The mortar batteries, below the camera, are not vis-
ible. (National Archives negative 111-SC-9795)

Fort Wool, 1984. In the foreground is a disappearing-gun position (note the counterweight well), part of the last Endicott-period work around Hampton Roads. In the left background stand the remains of the original Fort Calhoun, topped by a machine-gun position built during World War II.

Three-inch rapid-fire guns in the minefield-defense battery, Battery Irwin, during antiaircraft training classes at Fort Monroe, World War I
(1918). (National Archives negative 111-SC-9800)

Eight-inch rifle on railroad carriage, Fort Monroe, probably 1920s. (National Archives negative 111-SC-25510)

Standard 16-inch (naval model) emplacement of the 1920s. The gun had a 360-degree traverse but was vulnerable from above. This is half of Battery Murray, Panama Canal Zone, the first such emplacements constructed. (National Archives negative 111-SC-98611R)

The Submarine Mine Depot under construction, 1940. The Corps of Engineers was on its way to becoming the largest general construction organization in the world. (National Archives negative 77-DP-2375)

The ditch at Fort Monroe, 1984. A twentieth-century fire-control tower
stands in the center. Fort Monroe was the only active United States
Army post surrounded by a moat.

⊔⊓⊔⊓⊔⊓⊔⊓⊔⊓⊔⊓⊔⊓⊔ **4** ⊔⊓⊔⊓⊔⊓⊔⊓⊔⊓⊔⊓⊔⊓⊔

Interruption by War

Sieges, such as characterized the wars of the last century, are too slow for this period of the world . . . but earth forts, and especially field-works, will hereafter play an important part in wars. . . . It was one of Prof. Mahan's maxims that the spade was as useful in war as the musket, and to this I will add the axe.
—General William T. Sherman, 1875

The Third System had been developed to defend the United States against European attack, although without a lot of discussion about the likelihood of such an assault. Implicit in its history was another assumption—that a stout defensive system might deter a potential enemy from attacking. That point was mostly unspoken until Secretary of War Cameron viewed the forts as a guarantee against any European impulse to take advantage of the southern rebellion to prey upon the United States. That no European power landed its troops here (the French adventure in Mexico notwithstanding) owed more to other political, geographical, and economic conditions than to American fortifications, however. In any case, the Third System did not prevent the Civil War and had little real influence on its course. On the contrary, that war suggested a military irrelevance in the kinds of works the engineers had wrought over five decades.

The Civil War took control of American military policy—as expressed in the outlook of the army's leaders—away from the "scientific" experts of the Corps of Engineers and handed it to more ordinary mortals—to foot soldiers and factory hands, above all, and to officers of the fighting arms. Manpower and metallic devices, not architecture, spelled success or defeat in the field and in the war.

In retrospect, it might be suggested that the engineers could have predicted what would happen. But they were merely human, after all, so their academic assumptions about the nature of war blinded them to what was happening in real war, even though engineers—Delafield and George B. McClellan—were among the observers of the Crimean War. In that conflict, the Allies bypassed a fortified port to land on an open coast, then watched classical siege warfare deteriorate into general perniciousness under the conditions of mass warfare. Moreover, there was an iron-armored boat in the waters off Sebastopol, while on land other new weapons—especially the rifle-musket—wrought unprecedented carnage surpassed only by the toll of disease. The Crimean War was not scientific; it was ghastly. It was almost modern war.

The War at Hampton Roads

The presence of Union land and naval forces at Hampton Roads prevented the Confederates from using the area's fortifications for their intended purpose—to hold an attacking enemy off the coast and above all keep the foe from penetrating the heartland of Virginia. Fort Monroe could not be taken. Its strong works held a loyal Union garrison, and it was almost impossible to besiege or take by assault. Whether Fort Monroe could defend more than itself was not so certain at the outset, however. Its manpower was limited, although additional troops arrived as the crisis worsened.

The Gosport Navy Yard across Hampton Roads was of great concern, because it was the principal navy arsenal in the South. On 21 April 1861 Commanding General Winfield Scott ordered Captain Horatio G. Wright, "as an engineer officer of high science and judgment," to go to Gosport to help plan and execute the defense of the navy yard. Wright was too late to do much because the rebels around Norfolk were more than he could challenge. He took a new regiment just arrived at Old Point, ventured across to the navy yard, and destroyed all property that could not be removed. Wright and his men returned to Fort Monroe, and for the moment all the Union forces controlled was their own neighborhood.[1]

The Confederates built batteries at Sewells Point and other locations around Norfolk, including one at Fort Norfolk. By May 1861 they threatened to secure and fortify Newport News. Union general Benjamin Butler, the regional commander, put the neighborhood under military control and started active patrolling to keep the rebels at bay. Butler's men also strengthened advanced field positions. At the end of May he ordered a gun placed at the Rip Raps and later posted a garrison there. By September, Fort Calhoun boasted a 24-pounder on the wharf, seven 8-inch Columbiads in casemates, and two rifled 42-pounders awaiting carriages. It was thereafter secure as an advanced garrison and occasional home for political prisoners.[2]

There ensued a stalemate as both sides eyed Norfolk and the navy yard. Union strength built up steadily. The Confederates manned about a hundred cannons of various kinds, but they were not organized against a determined assault. Their hopes rested upon their small navy, secure in the Elizabeth River and formed around the *Virginia,* the ironclad rebirth of the *Merrimack,* a steam frigate scuttled at the navy yard by Wright's men. The ironclad abuilding at the Gosport was an open secret, and John Ericsson went to the Union Navy Department in September with plans for a vessel to match her. On 6 March 1862 the *Monitor* left New York for Hampton Roads.

The *Virginia* left the Elizabeth River while the *Monitor* was nearing the Virginia Capes. She was the product of no single genius but of characteristic Confederate improvisation. Engineers raised the scuttled *Merrimack,* pronounced the hull good, reduced the superstructure, and replaced it with a barnlike citadel of heavy timber clad in iron sheathing. The resulting monster was ungainly, unattractive, unseaworthy, and barely able to make headway, but there was nothing else like her in North America.

The *Virginia* made her debut on 8 March 1862, chugging ponderously into Hampton Roads. The Union fleet seemed helpless. The guns at Fort Calhoun fired in that place's first action against a floating enemy, but without effect. The *Virginia* went straight for the Yankee fleet. With their iron balls bouncing off her superstructure, she sank two warships and drove a steam frigate aground. Daylight and ammunition both ran short, so the *Virginia* retired into the Elizabeth River to replenish her stores. Her captain planned to finish the frigate and any other Union ships foolish enough to face him the next day. He did not observe the *Monitor* steaming into the roadstead.

Ericsson's invention was as monstrous as the Confederate creation. It was a long, low raft with only a foot or so of freeboard, crowned by a revolving iron turret mounting two 11-inch guns, with a smoke pipe aft and a pilot house forward. The *Monitor* greeted the *Virginia* in Hampton Roads on 9 March, and the two vessels hammered at each other for hours, until the *Virginia* retired up the Elizabeth River. In immediate terms the fight was inconclusive. The two ships never fought each other again, and the *Virginia* sank no more wooden vessels. When the Confederates retreated up the James in May, she drew too much water to go along and was scuttled for the last time. She and her opponent had changed the course of naval warfare, however. They threatened also to redress the balance between ships and fortifications. In firing on the *Virginia,* Fort Calhoun saluted the end of its own era.[3]

Before the two ironclads had their day, Benjamin Butler was replaced at Hampton Roads by Major General John E. Wool. Despite his advanced years and flagging energy, Wool soon had affairs well in hand. The United States government, after the *Virginia*'s appearance, wanted something done about Confederate Norfolk, and Wool was the man for the job. Secretary of War Edwin M. Stanton telegraphed him on 18 March "that in recognition of faithful service by a distinguished and gallant officer, the name of the fort on the Rip Raps is changed from Fort Calhoun to Fort Wool."[4]

President Lincoln went to Fort Wool on 8 May 1862 to watch Wool's mostly green troops attack Norfolk. The Confederate positions were not especially strong or well arranged, but neither was the Union assault. Nevertheless, the Federal forces had Norfolk in hand the next day. The rebels retreated up the James River, leaving behind the wreckage of military and naval properties—excepting Fort Norfolk, still intact, which returned to use as storage for the United States Navy.[5]

Norfolk was out of the war after May 1862. Union forces strengthened their grip on the area, building fieldworks at all avenues of approach. The Confederates made only one attempt to retake the place, when General James Longstreet tried to besiege Suffolk in April 1863. Union enigineers had surrounded the area with formidable entrenchments and strongpoints, and Longstreet declined to attack. He pulled out on 3 May, having failed in his main objective of drawing Union troops away from other areas.[6]

Hampton Roads became the base for the Army of the Potomac's campaign up the Peninsula during 1862, and throughout the war it was a strongpoint of the Union blockade of the Confederacy. Work continued throughout the war to strengthen the area's defenses, mostly by the excavation of earthen fieldworks and the placement of cannons everywhere. Fort Monroe brooded over it all as if in serene detachment. As for Fort Wool, its last guns were withdrawn by June 1864, although fifty-two casemates on its first tier were adjudged ready for armament. Construction of the original work resumed. Before long, however, cracks appeared in some casemate arches. Fort Calhoun's subsidence, thought to have ended, had returned to haunt Fort Wool. The project shut down once again.[7]

Every Man an Engineer

The conduct of the Civil War derived from the influence that army engineers—graduates of West Point on both sides—had attained in the military establishment in previous decades. Throughout the history of the Third System, the engineers sought to prepare the nation's coasts and harbors for modern war. The war that came upon them in 1861, however, was not quite what they had foreseen. Now the engineers must adapt to the brutal realities of real combat; in the process, their assumption that they were the army's finest brains was sorely tested.

They had been obsessed with fortifications since the birth of the Corps of Engineers. Now they must learn from trial and error in the field, on the coasts and inland. They also had an opportunity to test their theories and, afterwards, to refine their conceptions of the nation's defense and the engineers' role in it. How they met both challenges depended, to a great extent, upon how well they had learned their art from one man—Dennis Hart Mahan.

Mahan graduated first in the military academy class of 1824 and entered the Corps of Engineers. In 1832 he became West Point's professor of military and civil engineering. For the next four decades Mahan taught civil and military engineering and architecture, field and permanent fortifications, and the science of artillery. He also produced a body of works on which American engineering and much of American military theory were founded.[8] Mahan taught the leaders of both sides in the Civil War. The author of books on the "active defense" and the art of fieldworks made West Point's graduates engineers in outlook, whatever branch of the army they entered.

Union and Confederate soldiers were abundantly supplied with spades from the beginning and were inclined to use them. The deadly effects of rifle-muskets and explosive artillery shells drove men to the ground. Both sides learned by 1862 the superiority of the entrenched defense, and as a result when one side dug into the ground, so did the other. A war of maneuver repeatedly degenerated into a bloody war of local attrition, a contest for position between combinations of manpower and metallic devices.[9]

"At day break every body was 'up and doing,'" recalled a Union veteran of Chancellorsville in early May 1863. "It having . . . been decided that we should hold our position, large fatigue parties were detailed to clear a small space in front of our lines. . . . The trees were soon felled and distributed in such a manner as to seriously impede the progress of the enemy should they attempt to attack us." Union general Oliver Otis Howard said of that same day, "Our front was covered with rifle-pits and abatis." Howard was in for a rude shock. Stonewall Jackson's rebels struck from an unexpected direction, negating the products of hard work and less-than-scientific engineering and killing a lot of men in the process.[10]

Both armies in the Civil War pursued two kinds of engineering. One was military engineering proper, the construction of field-works and the like. The other was topographical, determining the lay of the land and the position of the enemy and setting the course of march. By 1863 both armies integrated the two activities into one organization. A new division in military engineering had developed by then—between the designated engineers and engineer troops on the one hand and the rest of the soldiers on the other. Engineer troops performed the most sophisticated bridge and field fortification work and topographical duties.

Disciples of Mahan, the officers, knew the importance of the spade on the battlefield. Before long, their men were equally oriented toward the ground, seeking protection from the murderous firepower carried by either side. Army regulations said that an order to dig in could come only from a division commander, but by the end of 1863 standing orders usually were to entrench whenever halted. Troops potted the ground wherever they went. They carried the art of field fortification to great heights, counseled by professional engineers but equally following their own experience and instincts.[11] Every man was, more or less, an engineer. General William T. Sherman found that not an unmixed blessing:

The habit of entrenching certainly does have the effect of making new troops timid. When a line of battle is once covered by a good parapet, made by the engineers or by the labor of the men themselves, it does require an effort to make them leave it in the face of danger; but when the enemy is intrenched, it becomes absolutely necessary to permit each brigade and division of the troops immediately opposed to throw up a corresponding trench for their own protection in case of a sudden sally. We invariably did this in all our recent campaigns, and it had no ill effect, though sometimes our troops were a little too slow in leaving their well-covered lines to assail the enemy in position or on retreat. Even our skirmishers were in the habit of rolling logs together, or of making a lunette of rails, with dirt in the front, to cover their bodies. . . . On the "defensive," there is no doubt of the propriety of fortifying.[12]

The Union and Confederate armies were at the mercy of new forces taking over warfare and beyond the ability of most of their leaders—especially their early leaders—to understand. The growth of modern economies and populations resulted in armed forces massive beyond precedent and equipped with equally unprecedented firepower. Engineering, which dominated military thinking, was an old-fashioned answer to an entirely new situation, and it did not work well. Seacoast forts could not protect their dependencies from new technologies, and field engineering merely added to the carnage—as illustrated in the futility of the Union campaign up the Peninsula in 1862[13] and the attrition of the campaigns up the James and through the siege of Petersburg later.[14] That the North finally wore the South down in the latter campaign was owing not to superior engineering or tactics but to economic superiority—the ability of Union leaders to marshal and manage industrial and transportation resources, especially railroads,[15] that exceeded the capacities of the other side.

Surveying the Ruins

If American military engineering—not to mention the Third System—needed further insult during the Civil War, it appeared in the campaign against Charleston, South Carolina, in 1863. There a fine harbor was defended by fine fortifications, with

Fort Sumter at their heart. Because it was the starting point of the Civil War, the Federal government wanted it taken. The navy tried first, sending a fleet of ironclad warships to wreck the forts and seize the harbors. The ships carried heavier guns than the Confederates had on shore, but the latter had more guns. Neither side could do serious damage to the other, and the result was a draw. Moreover, the ships were more confounded by fixed submarine mines ("torpedoes") and other obstructions in the channel than they were by the forts and guns. The latter, in fact, mostly protected the former, which did the real work of holding the attacker at bay.

The army was called in to silence the shore batteries, so the navy could clear the channel and resume its assault. A Corps of Engineers veteran and fortification expert, Quincy A. Gillmore, was in charge. The troops landed on Morris Island and quickly mired down in a siege against Battery Wagner, which had to be taken so that Fort Sumter could come under attack. The northern forces eventually prevailed against Battery Wagner, mostly by pouring more men and material onto the island than the southerners could match. The toll on both sides was ghastly. Gillmore then landed heavy rifled cannons and unleashed them on Fort Sumter, about a mile away. One of the Third System's finer monuments was blasted into rubble wherein not a gun could be mounted.[16]

The object of that brutish activity was not at all clear. If the government's desire was to demolish Fort Sumter for symbolic reasons, then the end was served—but at the cost of demonstrating the vulnerability of other Third System forts to new weaponry. If it was to take Charleston, then the campaign was a failure, because the city held out. Not even the threat of incendiary bombardment of its civilian population brought about a surrender, never mind the demolition of its fortifications. Charleston, buttressed by earthworks, was as secure without masonry forts as it had been with them. Even the debris that had been Fort Sumter declined to raise the white flag.

The record of the Charleston campaign offered many conclusions for future defense policy, if anyone had been inclined to read them. Ironclad warships were too new and unreliable to destroy a Third System fort; they might, however, become better. A Third System fort was too old-fashioned to destroy the new ships, even in their primitive state. The new rifled guns, however, could destroy a big fort, and maybe an ironclad, but first must be in position to do so. Engineering could neither protect nor conquer a port like Charleston on its own. And most curious of all, the ships were kept out of the harbor not by forts and guns but by simple obstructions in the water.

The Third System—and also the Corps of Engineers—had been born of age-old technical verities, eternal things such as geometry and earth and stone, with fire and water for dramatic effect and the science of ballistics for gentility. Now there was a new force in warfare, the combined genius of heavy industry and private invention. Machines and other sophisticated metallic appliances were intruding into the affairs of military men, and the soldier's life was no longer as simple as it had been when men built forts and other men attacked forts, and which side prevailed depended upon numbers, will, and time. Men still might build forts, but machines could destroy forts—and men built the machines.

Where did the men of the Corps of Engineers fit into the uncertain world of military affairs in the future? They had already been embarrassed by the mere scale of the Civil War, which enlisted scores of engineers from civilian life. Those newcomers proved adept enough at fieldworks, bridge building, and other activities long regarded as the special province of the trained military engineer. Even the common soldier in the ranks, a civilian in uniform for the most part, mastered the essentials of engineering his way into cover.

Moreover, civilian engineers in uniform dominated the new technologies that were becoming an essential part of warfare. The builders and managers of the United States Military Railroad, for instance, were not as a rule graduates of West Point, and few had any military background at all. They were experienced engineers nevertheless. That so many talented engineers were available was the bright side—it testified to the growth of the profession under the leadership of Dennis Hart Mahan and the United States Military Academy. Never again must the United States look to foreigners for engineering talent.

After the war, however, the Army Corps of Engineers would again be the exclusive province of West Pointers. Mahan had taught them to be flexible and creative, and the war forced them to adapt to many new things—railroads, aerial balloons, telegraphy, photography, rifled guns. They discovered new forms of transportation and rediscovered the virtues of earth as a fortification material. But they also saw an ironclad ship ignore a fort's guns as it attacked vessels under the fort's protection. They had seen—but had they understood?

Fighting soldiers in general had learned that engineering existed to support the army. But the Corps of Engineers collectively had spent its career on the Third System, seemingly with the assumption that the army existed to serve works of engineering. Individuals could understand that many things had changed— that the Third System no longer answered the national defense need, that many other factors besides engineering now figured in the military equation. Organizations, however, and especially ones with strong traditions and long-consistent outlooks, do not change as readily as individuals.

Wars interrupt the routine that is the characteristic and binding force of bureaucratic organizations, military or otherwise. The Civil War might also have been a good teacher, as good as Mahan. But the Corps of Engineers was six decades old, more accustomed to lecturing and doing than to learning. To anyone outside the corps, one lesson of the war was clear—military engineering must change, and so must the outlook of the Corps of Engineers. Within the military engineering fraternity, however, that message was not so apparent.

*The engineer has solved, with mathematical certainty,
the problem*

⊔⊓⊔⊓⊔⊓⊔⊓⊏ **5** ⊔⊓⊔⊓⊔⊓⊔⊓⊏

The Struggle to Preserve the Old Order

"Hold the fort, for I am coming,"
Jesus signals still.
Wave the answer back to heaven,
"By the grace, we will."
—Philip Paul Bliss, 1870

It appeared for a considerable time the Corps of Engineers had learned very little new about military defense from its experience in the Civil War, no matter how proficient many of its members had been. In following years the organization was increasingly occupied with river-and-harbor improvements and related civil works. At its heart, however, the corps conceived itself still a military organization and the intellectual heart and soul of the army.

Few others in the War Department shared the corps's vision of itself. Its self-assumed elitism bore little weight among soldiers who believed the fighting arms had achieved the victory over the rebellious South. The engineers, accordingly, in 1867 lost the control over West Point that they had held since the founding of the military academy, and as a result the outlook of the army as a whole increasingly diverged from that of the Corps of Engineers.

At the top of the corps were men who had performed capably in the war but had reached professional maturity during the prewar days when the engineers represented the army's best thinking about modern warfare. That had also been the time when their first mission was the fortification of ports and harbors according to principles laid down by their founders. Younger men of more current vision were rising in the organization, but the leadership—distracted as it might be by civil works—remained devoted to the forts. Thus was waged a last campaign to perfect the Third System. It was a lost cause.

Holding the Forts

Chief engineer Delafield concluded that the Civil War had taught three lessons: (1) fort architecture must be revised becaused masonry was vulnerable to rifled guns; (2) mining waterways with fixed torpedoes was an essential part of future defense; and (3) the United States needed a larger professional army. In other words, defense in the new age would require more of what had gone before, with minor modifications.[1]

The chief appointed two boards in June 1865 to consider fort modifications to accommodate the largest guns and to permit masonry walls to withstand rifled shells. Fort Monroe was the site of experiments on iron armor to protect guns and gunners; the first tests were inconclusive. The chief called for more experiments to determine the power of modern weapons and the value of wrought-iron plates for scarp walls. With iron plating he hoped to adhere to the lines laid down by Simon Bernard a half century before.[2]

The engineers were not alone in their reluctance to cast off old patterns. The navy also was burdened by custom and inertia. An army-and-navy Harbor Defense Board assembled in 1866 to consider defense needs and whether the army or the navy should handle them. The board identified separate missions and divided into subcommittees to address them independently. They decided that some new system was necessary, involving both strengthened forts and ironclad ships. As in the past, the navy would be the first line of defense, the forts second. Ships and forts alike, save for being covered with iron, would be and perform much as always. The board never met again.[3]

The engineers returned to their forts, shoring them up to carry bigger guns and carriages, which except for size and weight were like what the forts had always housed. Experiments with iron shielding continued; the results remained inconclusive. Appropriations declined, so far that by 1867 the engineers were down mostly to maintenance of ancillary structures and plugging holes caused by riflery during the war. The chief of engineers formed another board of officers in 1867, telling them to determine the force of the largest artillery, means of using it in the forts, and ways to protect masonry forts and their occupants. The results, he said a year later, "have not yet afforded all the desired conclusions." Accordingly, he planned to continue in the pattern so long established, giving only a hint of the future when he mentioned that disappearing carriages for barbette guns merited consideration. Meanwhile, he complained of increasing resistance to fort appropriations.

Congress, in fact, was well on its way to a quarter century of downward pressure on military budgets that held the army to a bare minimum. The trend began with postwar reaction and legislative-executive rivalry growing out of Reconstruction, but at its heart lay persistent American antimilitarism compounded by doubts that the fort builders knew how to deal with rapid technological progress in gunnery, explosives chemistry, metallurgy, ship design, and tactics. Nor were many in Congress persuaded that the country faced a real threat from abroad.

The chief of engineers announced in 1869 the results of tests at Forts Monroe and Delaware. The engineers had built experimental walls and shot directly at them. They concluded that iron would be effective as casemate shielding, but it was too expensive for widespread use. The engineers nevertheless prescribed the future of fortifications as they viewed it: (1) barbette batteries of very large guns were needed, along with magazines and bombproofs; (2) a depressing gun carriage ought to be developed as a substitute for barbette carriages currently in use; (3) the use of many large mortars in defensive arrangements would be advisable, to puncture the decks of ships; (4) torpedoes should be adopted as accessories to forts; and (5) obstructions and floating batteries also should be adopted as accessories. The Corps of Engineers, the chief announced, would develop proposals for specific projects following those prescriptions.[4]

Modest appropriations supported fortification work until 1875. Although the engineers mostly rehabilitated established structures, there was planning for new construction, which revealed some advanced thinking. A new trend eschewed large, concentrated structures in favor of separated earthen batteries, which offered advantages in cost and tactical effectiveness. The chief of engineers concluded by the early 1870s that harbor defense was solely the province of the engineers, sailors and others need not apply. He went beyond forts only so far as to call for mining harbors at the outbreak of war and mobilizing a fleet of torpedo boats. The forts would be manned by trained artillerists but all else by combat engineers.[5]

Appropriations were so low by 1870, however, that the engineers could do little more than the most fundamental maintenance, along with continued studies of iron shielding. The chief announced in 1871 that casemated forts would always be necessary and in 1872 declared that torpedoes, boats, and obstructions were mere accessories requiring protection by the forts to which they were attached. The only significant new development in design was a trend toward high parapets and guns on descending carriages. Congress granted the first money to acquire torpedoes for harbor defense on 21 February 1873, catching the engineers unprepared; a delegation went to Europe to study torpedo systems. Meanwhile, structural studies went far enough by 1874 for the engineers to announce a sweeping plan to fortify harbors deep enough to carry enemy ironclads. A new series of forts, designed on traditional principles, would defend against naval attacks. Their most distinguishing new feature was that they would be mostly earthen barbette and mortar batteries with "great thickness and height of parapets" and "massive traverses and parades." The works should be designed for gun carriages of "increased height" and would also serve for "depressing-carriages when that shall have been provided." Experiments on iron and torpedoes would continue.[6]

The future of American coastal defense was problematic amid a swirl of progress in metallurgy and weapons design. Few people outside the Corps of Engineers believed that further investment in old-fashioned forts was wise, while the engineers all but ignored the ideas of others. Congress gave the forts their last appropriation on 10 February 1875, and within a year work on them almost ended. Chief of Engineers A. A. Humphreys made ever stronger arguments to save the program, taking his case to Congress. There was danger of enemy raids from Caribbean ports, he said, and we had nothing but neglected works to face them. Increasingly, Humphreys sounded like his predecessor Totten in his arguments to get old works upgraded and new ones built before hostilities commenced. The principal innovation Humphreys supported was lines of fixed, electrically fired torpedoes, which he explained would obstruct harbor entrances, holding enemy ships under the fire of shore batteries. Electrical equipment could not be obtained quickly in an emergency, he said, so work on torpedoes should begin immediately. In 1878 he offered a ringing tribute to casemated forts, which the engineers had decided to secure with wrought-iron plating. Perils were at every hand, he declaimed, and the European powers were spending far more than the Americans. He offered plans for iron-plating three old forts at New York as a demonstration of what he hoped to do.

Money was not available for anything more than maintenance, however. The engineers occupied themselves with plans to upgrade the old forts, which by 1879 produced a new proposal for Fort Wool, among others. By the 1880s the engineers turned increasingly to historical examples to justify resuming work, purportedly demonstrating the value of fortifications and the perils of not having them. The costs of an enemy raid on a major city, for example, were compared to those of the Chicago fire of 1871.[7]

The military appropriations act of 3 March 1879 restricted operations at coastal forts to protection, preservation, and repair. Chief Horatio G. Wright issued another declamation on the threat of attack, the vulnerability of American ports, and the costs of a bombardment. The navy alone was not enough to defend the country, he said, because the enemy could pick its landing point. All ports and harbors must be guarded with modernized works, and so should all landing places: "We can make but a feeble defense against the fleets now prepared and rapidly increasing which will sooner or later be brought against us by some of the most powerful nations on the earth, or by others nearer at hand whose offensive naval means exceed our own, and whose powers are not to be despised."[8]

This sort of hysterical extravagance became self-serving when it was joined in 1881 with a declaration that forts cost less than battleships. Congress refused to approve anything new for the old forts, however. In 1884 Chief of Engineers John Newton tried a new tack. England, he said, was the threat. However, if appropriations resumed, the engineers could face down the mighty British: "For the first time in the development of the modern art of war, the engineer has solved, with mathematical certainty, the problem of closing harbors and rivers against hostile ships, so that the sole question in each particular case would be whether the importance of the place would justify the cost."[9]

Congressional reluctance to renew fortifications work proved fortunate. Repeated revisions of plans ultimately turned the engineers' eyes from the past to the future. The fortification question was swept up in a larger movement to upgrade the whole army, which in the 1880s was scattered around the country in small posts, little more advanced militarily than the Indians it guarded. Under the leadership of Commanding General William T. Sherman, the War Department launched a program to consolidate the army, reorient it toward the coastline, and modernize the force, its equipment, and its doctrines. Part of the program was the assembly of a board chaired by the secretary of war, which in 1885 gathered the best experts—not engineers alone—to devise a new system for harbor defense. The resulting plan occasioned the practical abandonment of three generations of coastal forts.

The engineers, however, continued for some years to promote the older works. That determination sometimes interrupted funding, as in 1886 when Congress withheld money until the engineers could decide what ought to be done. "It is believed," said Newton, "that the failure of Congress to make appropriations for this all-important work has mainly arisen from the difficulty in determining the best method of procuring armor-plate." But Congress could not be persuaded that further investment in the old places was justified and into 1887 did not even provide for maintenance. Chief of Engineers James C. Duane protested that "economy requires that they should be kept in repair." His successor Thomas L. Casey deplored the "additional damage and deterioration incident to over two years of absolute neglect." So long as Congress hesitated to start new construction, the engineers remained devoted to their old creations. Then budgetary penury took the forts from them. "The failure to obtain any appropriations for 1887," Casey said in 1888, "necessitated the practical abandonment of all permanent and other defenses where there were no garrisons, or where ordnance sergeants could not be detailed to take charge. Portable property was secured as well as possible and the fort-keepers discharged."[10]

Small appropriations resumed for maintenance and preservation in 1889 and 1890, but that work was soon overtaken by construction of entirely new defenses. The engineers tried, almost sentimentally, to integrate the old works into their new plans, but that effort was mostly futile. Nevertheless, even in the 1890s some engineers refused to abandon the past. Lieutenant Colonel William R. King, for instance, looked with disfavor on the newest developments, scorned the "quick-firing gun," and could see no advantages in smokeless powder. Muzzle-loaders, he declared, were still useful. His fort of the future resembled something out of the War of 1812, with earthworks and iron guns.[11]

Maintenance appropriations ebbed as swiftly as they resumed, while the more forward-looking engineers grew involved with modern works. The inspector general's office studied twenty-seven abandoned coastal forts and decried their neglect. "As matters stand," said senior Inspector General Joseph G. Breckinridge in 1893, "the forts are going to ruin, and the condition of their guns, carriages, and ordnance stores generally is reported as disgraceful, notwithstanding they are included in the defensive plans of the seacoast, more especially for the operation and protection of torpedo lines. Can this condition be continued without any effort to improve it?" Commanding General John M. Schofield echoed him: "Many of these posts have now become defenseless and unfit to occupy in their present conditions."[12] It was no use. Appropriations for preservation in the 1890s seldom exceeded $45,000 a year. Despite occasional assertions that the old forts were still useful, the engineers now were too distracted by the new works. By 1899 the new structures absorbed most of the money available for preservation and maintenance, and the Third System forts faded away.[13]

The Corps of Engineers meanwhile lost an opportunity to become a general construction agency for the army. That the Quartermaster Department would continue to control the majority of army construction was confirmed in the resolution of an old problem—the housing of men in casemates. The engineers had since the 1820s confined themselves to "fortifications" and left ancillary buildings to the quartermasters. The latter provided all the army's housing and other nonfortification facilities and before the 1880s did most of its work west of the Mississippi. The first efforts to reorient the army to the seacoast revived the casemate barracks issue. The surgeon general's office surveyed army housing in 1870 and 1875 and issued scathing indictments of living conditions almost everywhere, but particularly in casemates. The work of the surgeons encouraged many reforms over the following years, except at the coastal forts. So long as most of the army was in the West, that was a small problem. When garrisons moved to the coasts in the early 1880s, the issue arose with a fury.[14]

The Corps of Engineers held firm that quarters were the responsibility of others—or equally, that quarters were not needed at coastal forts. Many in the army thought otherwise. An infantry officer, for instance, hoped in 1881 that the engineers would take a different view of the question: "Our Engineer Department will not, so far as can now be foreseen, recommend to the Secretary of War, that any attempt be made to provide quarters for the occupation in time of peace, of the garrisons of permanent works of defense yet to be erected, when there is room for such quarters on the exterior. Casemates are now called war quarters by the engineers, and their use in time of peace as quarters for either officers or men, will doubtless be given up as soon as it can be done."[15]

Solving the barracks problem depended on Congress's willingness to pay for post consolidation and new construction. That began in the mid-1880s and accelerated in the 1890s, but before it was felt on the seacoasts, the housing situation there became increasingly serious. "The cellar-like condition of any and all casemates," said an inspector general in 1889, "renders them totally and absolutely unfit for human habitation." He explained more fully the following year: "As a Christian nation, and as a humane people, we owe it to ourselves to do away with the thoughtless cruelty that is being practiced on the officers and men who are compelled to accept this sort of shelter. . . . To enter these dark, damp, cellar-like quarters on a hot day in summer and find yourself confronted by a large base-burner coal-stove in full operation gives you both a mental and physical shock. The nation's most vicious criminals are provided with better quarters than are given these officers and men of the Army."[16]

The quartermaster general asked for special appropriations so that his department could build quarters at coastal forts. "It is not humane to make men live in these structures which are unhealthy, unsuitable, and injurious to the mental and physical character of the occupant, except when in time of war great necessity forces it," he said. The quartermasters had already begun to act, however, and they did so at Fort Monroe, pride of the Third System and one of the few places where the engineers had built some proper quarters. Those were old and insufficient by 1880. In 1881 the quartermasters delivered a prototype for a new generation of army housing. It was a large, two-story building housing six companies, affording airy living space and modern plumbing. The post commander was delighted, especially with "the entire absence of the usual foul odors of enlisted men's quarters."[17]

The engineers no longer dominated the landscape. The quartermasters did more construction at Fort Monroe than the engineers ever dreamed would take place and blurred traces of Bernard's design. The engineers remained active at Fort Monroe, of course, because its vicinity hosted some of the first of the new defenses. Their construction also disturbed the original design, because the conception of the place as a bulwark faded away. When that happened, the engineers surrendered even maintenance to the quartermasters.

The final solution to the casemate-housing problem developed in 1902 and further eroded the engineers' role in fortification. The secretary of war declared that the real source of the difficulty lay in the "false economy" of buying just enough land for the forts and none for other facilities. Therefore, the secretary ordered that proposals to buy land for coast defense must be certified by the chief of artillery to include enough land to meet all needs. He did not mention the Corps of Engineers, and need not, for its role had been reduced considerably. Engineering had taken second place to gunnery, in a process that began when the first rifled shell broke open a fort wall.[18]

Holding Fort Monroe

The first postwar work involved two efforts at Fort Monroe. One was renewed construction and upgrading of the facility to mount superheavy guns. That terminated after 1875. More enduring were maintenance and preservation. This activity often resembled urban design more than military engineering, and it declined steadily when the engineers took up new works. By the twentieth century they had moved to the waterfront, leaving Fort Monroe to its commander and quartermaster.

Construction resumed in 1866 under the direction of Colonel Henry Brewerton of the fortifications office in New York. His crews reinforced the barbette gun platforms, installed front-pintle platforms for 15-inch guns, and adapted the water battery for new 10-inch guns, using concrete paving for the first time. Maintenance work included repairs to pavement, grading, and masonry. The routine continued. In 1867 the engineers reconstructed gun platforms and repaired masonry and ancillary structures and bridges. By 1868 the work load expanded considerably, although it was mostly of the same order. Maintenance included demolition and reconstruction of the breast-height wall and parts of the water battery. Late in 1867 Brewerton took up Fort Monroe's most frustrating project, the completion of the artesian well started in the 1840s. He had to abandon the original well and start a new one. By the end of 1868 the hole was down 370 feet.[19]

Brewerton retired in 1867 but continued on the job until transferring it to Major William P. Craighill of the corps's Baltimore office in 1869. Additional platforms for 15-inch guns appeared during his last year, along with the usual maintenance and repairs, and some obsolete wooden structures were removed; but tightening of the budget began to cut into the work. As for the well, Brewerton pressed ahead, sinking 8-inch pipe until the lowest section separated at the 575-foot level. Rather than draw it out again, the engineers inserted 585 feet of 5½-inch pipe into it. The hole passed out of clay at 574 feet, then into a clay-sand mixture, and finally hit a saltwater layer that gushed to 4½ feet above ground level in twenty-four hours. Brewerton remained optimistic. "It is proposed to continue the operations of the well as long as a reasonable prospect exists of reaching pure water within a moderate depth," he said.[20]

Construction appropriations expired in 1870, leaving the engineers little to do except prepare and revise plans. Operations focused on repairs and on driving the well down to 900 feet. There was money only for the well the next year, but its auger wore out at 906 feet and the work stopped. The well drilling resumed on 1 October 1871, and a few repairs to the fort followed in 1872. A geologist assured Craighill that he should strike water at 1,200 feet down, so he aimed at that level.

In fact, however, the work soon stopped for want of money and because of its increasing difficulty. Craighill paused to restudy, while his officers resumed repairs and started modifying the redoubt north of the main fort according to a new plan. The Board of Engineers for Fortifications also submitted a plan for "barbette-batteries at Fort Monroe" using King's depressing gun carriage, an experimental model. Craighill's forces forged ahead in fiscal 1874, building new gun emplacements and overhauling the redoubt for its new armament. They also studied the possibility of plating the main walls with iron and laid a railroad track to carry materials. Momentarily setting the well aside, the engineers built a cistern instead.

It appeared that Fort Monroe soon would be thoroughly upgraded. Craighill prepared, and the secretary of war approved, a plan "for a battery of ten guns of the heaviest caliber, exterior to Fronts II and III"; although a "project for a continuous battery on fronts I, II, III, and IV of the main work" was rejected, "a new project for a battery of two guns has been made and approved." The first project would convert the water battery from casemates to a barbette battery. On 8 August 1874 Craighill transferred Fort Monroe to Lieutenant Colonel Quincy A. Gillmore—of the Charleston campaign—who managed it and the rivers and harbors of South Carolina, Georgia, and the Atlantic coast of Florida until his death in 1888. But Fort Monroe faced hard times, because appropriations ceased. Most work in 1875 was on rebuilding the redoubt, and Gillmore laid plans for further construction and reopening of the well. By 1876 he had part of the battery in the redoubt almost ready and offered three heavy-gun platforms and fifteen light-gun platforms to the artillery, but the structures sat empty. By 1877 the work was down to the most necessary preservation, except for dismounting four old 10-inch Rodman guns, dismounting and remounting three 100-pounder Parrott rifles, and mounting four 8-inch converted rifle-guns. Only "slight repairs" to the breakwater and a timber platform for a 13-inch seacoast mortar were possible in 1878, and not even that much the next year.[21]

Gillmore completed in 1880 the last major plan to retain the Fort Monroe of old as part of the seacoast defenses. It called for arming the redoubt with heavy barbette guns, supported by traverses and traverse magazines; similar armament in the reentering place-of-arms, on the right of the redoubt before front five; building a new barbette battery to the right of the water battery; and putting a few heavy guns in the salients of the main work and covert way. "There is plenty of room for mortar batteries both within and without the work," he added. In fact, nothing like that would take place, and eventually the old fortifications would give way to something modern. Work had been suspended for several years. By 1880 the two traverse magazines of the earthen redoubt stood finished but uncovered by earth, and six gun platforms were "nearly completed." Most other alterations were in lesser states of completion, and the ten-gun earthen battery had not even been started.[22]

A few minor repairs were made in the following years, especially to eroding embankments, and Gillmore asked futilely for money to do more. In 1883 he took up the railroad track and stored it, put housing over uncompleted structures to preserve them, and repaired a breakwater damaged in a storm. The next year he started building a sewer main and laterals. The Corps of Engineers recognized by that time that Fort Monroe was outdated and should be reexamined. Something more than larger armament was in order. "It is expected that the changes finally adopted will be of a much radical character" than those planned in the 1870s, said Chief of Engineers Newton. Meanwhile, the engineers made only the most essential minor repairs.[23]

The engineers began to pull out of Fort Monroe during the 1880s. The post quartermaster took charge of maintenance, except for those things that were strictly classed as fortifications. A board of officers divided land in the reservation to "consolidate separately, the belongings of the Engineer Corps and the Ordnance Department." The government had decided to build radically new defensive works, most of them outside the boundaries of Fort Monroe proper.

Congress, however, now decided not to allow the engineers to leave the place altogether. It appropriated funds by 1887 to build a new wharf and upgrade the road leading to it. Lieutenant Colonel Peter C. Hains designed a project, and in October 1887 a contractor started to build an iron wharf structure. Hired crews, meanwhile, widened the roadway from the wharf to the Mill Creek bridge. On 10 August 1888 Congress authorized expansion of the wharf project; Hains redesigned the structure to be longer and wider than originally conceived, with an additional fender system. Then on 2 March 1889 Congress granted money for an iron bridge over Mill Creek. The new wharf was opened to public use in September 1889 and the new bridge in May 1890.[24]

The wharf and bridge were not authorized for a military purpose. Instead, the Fort Monroe waterfront hosted several famous hotels, and Congress wanted to give pleasure-seekers better access. Accordingly, when the lawmakers appropriated funds to start construction on new defenses in the 1890s, including some for Hampton Roads, they added money to protect the shoreline at Fort Monroe and for the defunct artesian well. During the time that the engineers embarked on these nonmilitary projects at Fort Monroe, they were told to leave instructions and experiments in submarine mining there to the artillery school.[25]

By fiscal 1891 the engineers had separate projects at Fort Monroe for beach protection, development of a water supply, and construction of a sewerage system adopted in the act of 2 March 1889. Erosion of the beach was the most immediate matter. Northeast of the fort the water threatened to cut through the strip connecting Fort Monroe to the mainland. Hains contracted on 4 March 1891 for a series of timber groins; and even before they were completed, the beach began to restore itself. As for the water supply, Hains wanted to put the first $10,000 into the artesian well, aiming now at 2,500 feet into the ground. He received only $6,000, which he regarded as not enough to start work. He asked for the additional $4,000, and his successor raised that figure to $14,000 in 1893. Work on the well did not resume.

That left the sewerage system, which Congress wanted complete and integrated inside and outside the fort, serving the government and civilians alike. The first appropriation did not cover the first bids, however. In February 1891 the engineers inspected the private systems on the military reservation, decided they could be connected to a general system with modifications, and revised the plan to meet the limits of the appropriation. In October 1891 the secretary of war decided that two systems must be built—one by the government and the other by the hotels and other private parties. A revised estimate went to Congress in March 1892, along with the testimony of the post surgeon regarding "the greatest urgent need of this improvement." Fear of a cholera epidemic soon produced "a very voluminous official correspondence" on the need for decent sewerage. In September 1892 an alarmed secretary of war, goaded by the surgeon general, called for one entire system. The engineers had plans ready by October. Finally, on 1 August 1894 Congress appropriated one-half the cost of a sewerage system for all buildings on the Fort Monroe reservation; the private parties involved would pay the other half. A contract was let in January 1895 to provide a complete system— pipe, manholes, concrete, and all. It was in operation, after several delays, by early 1896 and was turned over to the post quartermaster on 1 September.[26]

Lieutenant George A. Zinn made "very extensive repairs" to the Fort Monroe breakwater in 1895. It had washed out for over 60 feet, and the bank behind it had been cut back 10 to 40 feet, raising a danger that the sea could cut through the moat. Zinn built a concrete wall and additional groins to protect the beaches northeast of the fort. His successor built more concrete breakwater between the Hygeia Hotel and the Engineers Wharf and in following years continued to replace old piling with concrete. Meanwhile, Captain Thomas L. Casey, Jr., as frustrated as his predecessors that no more money had appeared for the artesian well, tried to contract for that project, but the bids were too high.[27]

For the first time in several years, the engineers did some minor repairs around Fort Monroe itself in 1898 and prepared a new map of the military reservation. However, Fort Monroe looked more and more like a civilian wateringhole and less and less a military reservation. The last thing the corps did there as a project of military engineering, so called, came in 1904, when it reviewed plans of the Hampton Roads Railway and Electrical Company for the another bridge over Mill Creek. Fort Monroe remained, but the Third System—of which it was the grandest monument—was no more. Gone also from its gates was the Corps of Engineers, which had new responsibilities on the waterfront.[28]

The Decline of Fort Wool

Work resumed at Fort Wool in 1866. Brewerton had crews lay masonry and install embrasure irons and flooring on the second tier. They also rearranged the loading that held the place together. He decided that the structure's foundation problems were correctable and opened the old foundation up and filled its spaces with mortar. Then he tried to fill voids in the "encroachment" surrounding the foundation, pouring in mortar for grout. The area swallowed tons of sand and gravel and was still full of spaces.[29]

Fort Wool stood on literally shaky ground, was exposed to foul weather, and the engineers had limited time for it. Nevertheless, they inched ahead on the second tier and in 1867 began to build magazines for the first tier of casemates. Constantly they piled and removed stone to stabilize the place. By 1868 they concentrated on completing the first tier and all its "adjuncts," setting aside the goal of a three-tier castle on the Rip Raps. They continued work until 1870, when the chief of engineers stopped the job with fifty-two casemates ready to receive guns.[30]

Until the engineers had new plans to complete Fort Wool, they decided to do no more than preserve what they had produced so far. By the end of 1870 they devised a scheme to upgrade the place. Rather than continue with the second and third tiers, they would substitute a barbette tier on top of the first, to mount 12-inch rifled guns. The casemates would be reserved for 9-inch rifles. Nothing came of that plan, however. It was discarded by 1872, and the question of a new version was open.

Apparently the engineers hardly visited the place for several years; the seabirds and a solitary fort keeper had Fort Wool to themselves until 1878. That year, Professor W. K. Brooks of Johns Hopkins University arrived with a colleague and eight students and turned the fort into a laboratory. They spent the summer conducting pioneering research on the invertebrate zoology of the lower Chesapeake Bay.[31]

Brooks's fishing expedition may have reminded the engineers of their unfinished business on the Rip Raps. They returned to repair the wharf and the fort keeper's quarters and drew up new plans. "The plans contemplate an armament of the heaviest modern rifled guns behind iron shields, and an appropriation for the work is urgently recommended," said the chief of engineers. The proposal would arm part of Fort Wool with thirteen 12-inch rifles in casemates behind the shields, along with seventeen 8-inch rifles. These specifications changed quickly as the Board of Engineers spent most of fiscal 1879 refining the scheme. The members first returned to the discarded plan of 1870, calling for a barbette tier for 12-inch rifled guns, with 9-inch rifles in the casemates. They soon changed that to twenty-one 81-ton guns and five 12-inch rifles and declared this plan to be the first based on the new technologies. It proved too costly, however, so the board looked for something less expensive, submitting a final report on 14 June 1879 calling for a considerably smaller development. "In connection with the works devised already and that will be required in the future on the Fort Monroe side of the entrance to Hampton

Roads," said the chief of engineers, "the less costly plans [for Fort Wool] are deemed sufficient."[32]

The chief requested appropriations for the new project, but he received none. The biologists did not return in 1880, and neither did the engineers. The chief proposed "to modify this important work so that it may receive the heaviest modern rifled guns, protected by impenetrable iron armor. The work to be done, which will require several years for its execution, cannot be left until the near approach of war, and a liberal appropriation for it is urgently recommended." He again asked for money to start the work, but he never saw it. In 1883 crews repaired and painted the fort keeper's quarters and other buildings and repointed part of the masonry of the second tier of casemates. More minor repairs followed in 1884 and 1885 and a new wharf in 1886, and that was the end. Fort Wool all but dropped from sight, with the government's only interest in the place $600 a year to pay the fort keeper. Even that ended on the last day of 1900, because the appropriation was only $300 that fiscal year. Now only the birds ruled the castle on the Rip Raps.[33]

The Corps of Engineers had tried earnestly since the Civil War to hold on to Forts Monroe and Wool and the other cherished works of the Third System, but it was no use. Time had passed the old forts by. Before the fort keeper departed the Rip Raps, an entirely new generation of coast defenses was abuilding. And a new generation of engineers was working on them—no longer as the dominant force but in (not always congenial) concert with other experts who had shouldered their way into coast defense policy. The new works forced the engineers to change their outlook and their role in national defense, but they also captured the soul of the Corps of Engineers as thoroughly as had the Third System of yore.

⊔⊓⊔⊓⊔⊓⊔⊓⊔⊓⊔⊓⌐ **6** ⊔⊓⊔⊓⊔⊓⊔⊓⊔⊓⊔⊓⌐

The Decline of Military Engineering

At no stage of our national life have we been
brought face to face with the armed strength
of a great world power free to land sufficient
forces to gain a foothold at any desired portion of our coasts. That we have to some extent felt this danger is evidenced by our
efforts to provide a navy as a first line of
defense and to supplement it with the necessary harbor fortifications; but we have not
yet realized that our ultimate safeguard is an
adequate and well-organized mobile land
force. Experience in war has shown the need
of these three elements but the public has not
yet demanded that they be perfected, coordinated, and combined in one harmonious system of national defense. *Not until this has
been accomplished will a proper military policy for the United States be adopted.*
—The General Staff, 1915

The 1879 plan for Fort Wool was part of the last attempt to keep the fortifications as they always had been. Rejection
of the engineers' plan foretold a new approach to coastal defense,
amid one of the loudest public debates over military policy during
the nineteenth century. The discussions produced yet another
comprehensive system for the nation's coastal fortifications. The
engineers, however, no longer dominated official thinking as they
had earlier in the century but found themselves contending with a
congeries of new technologies and newly assertive, competing experts who had their own ideas on defensive matters. Implementation of the new plan further reduced the engineers' role in defense
planning until at the end the corps followed designs mostly dictated by the requirements of others. It was a grand experience,

however, and by the time it was over the corps was as heavily invested spiritually in the new defenses as it had been in the old. Unwilling to face termination of the program or the frightening new technologies that made it quickly obsolete, the corps looked again to the distant past, evoking the Third System as a stable refuge in a fast-changing world. As in the Civil War, however, new developments in military technology were inexorable, and they were most apparent around the Third System's heart, Hampton Roads.

The Endicott Period

Commanding General Sherman started the public debate in 1880 when he denied the assumptions underlying the engineers' program. "[M]ost of our sea-coast forts are superfluous," he said. "We now have fifty millions of people, and the idea of any hostile force landing on our coast is simply preposterous. . . . All minor forts should be abandoned." Sherman did want defenses at the major ports strengthened, but not by the corps alone. "Artillery officers," he continued, "should also be associated with the engineers in constructing, altering, and repairing seacoast forts, because the men who have to fight these batteries should have something to do in their construction."

Sherman did not believe harbor defense the most important defense issue, as the engineers maintained. Rather, reorganization of coastal defenses was one part of a reform of the military establishment focused on post consolidation, improved training and equipment, and above all modernization. Orienting the troops toward a civilized enemy off the coast would make them a truly modern fighting force. That goal was further reflected in his condemnation of the artillery in 1881. "I also invite attention," he said, "to the absurdity of styling in orders the companies of foot artillery, armed with muskets and without guns, 'batteries.' They are not batteries in any intelligent sense."[1]

Technological advances during the 1880s outdated the Third System and its latter-day defenders. New metals and manufacturing processes permitted large, rifled, breech-loading weapons of unprecedented range and striking power and ships impervious to old-fashioned smoothbores, no matter how heavy their projectiles. Other developments included the growing use of portland cement concrete in construction and advances in optics, explosives chemistry, and ballistics. Also important was the resurgence of the artillery, which shared with the Ordnance Department increasing scientific sophistication and the fervor of youngsters who saw themselves prophets of a new order.[2]

"Our seacoast, with its great cities and important harbors," Secretary of War Robert T. Lincoln declared in 1883, "is defenseless to-day against the attack of a modern iron-clad, and it is humiliating even to imagine the mortification, loss of life, property, and prestige to which we should be subjected should war come suddenly upon us, as, the history of nations shows, may happen at any time." The secretary's warning sounded like the voice of the engineers. As events demonstrated, however, he listened more closely to others.

The first response of Congress to the clamor for new coast defenses was a rider to the Naval Appropriations Act of 3 March 1883. It called for a board of six army and navy officers to consider setting up a government foundry "for the manufacture of heavy ordnance adapted to modern warfare, for the use of the Army and Navy of the United States." The Gun Foundry Board reported in 1884. It advised both a government foundry and market purchases to encourage domestic steel production, along with renewed fort construction. The board was echoed by Sherman's successor Philip Sheridan, who called attention to "the perfectly defenceless condition of our seaboard cities and their harbors against foreign naval attack."[3]

Frightening the citizenry did not bring immediate appropriations, because the War Department was not certain what it wanted to do. That was excepting only the engineers, who realized that leadership in the business could pass to others; they, accordingly, were the first to advance a comprehensive plan, offered in 1884. The philosophy of the Board of Engineers "congealed" into what was presented as a whole new system of coast defense design: (1) build lines of submarine mines or other obstructions to detain enemy vessels under fire; (2) defend those obstructions against small vessels; and (3) develop defenses to keep armored vessels from closing on the mined areas to silence their defenses (described as the most difficult and costly, not to mention vague, part of the system).

The board's ideas were merely the old fortifications with some up-to-date window dressing. The most striking—and impossibly expensive—part of the plan called for turrets and casemates protected with wrought-iron armor 36 inches thick. The board prepared estimates for heavy guns and rifled mortars, and their emplacements, for Hampton Roads and six other major ports. "Although these estimates are approximate," said the chief of engineers, "they have been made with care, and it is believed that the fortifications (including guns) of all harbors on the coast of sufficient importance to tempt an enemy can be prepared at a cost of not to exceed $60,000,000." He asked for money to start work at Boston, New York, Baltimore, Hampton Roads, and San Francisco and to restore the old Spanish work of Fort Marion at St. Augustine, Florida.[4]

That proposal brought no response, partly because the engineers no longer held a monopoly on harbor defense. Increasingly the artillerists made themselves heard. They at first sounded like the engineers, justifying a renewed program by the values at stake. Said Lieutenant G. N. Whistler in 1884: "However, it should be remembered that the peculiarities of the geographical position of our country are such that, save in a war with England or Mexico, our light artillery would have comparatively little to do in foreign war: and that, in any such war, the more scientific branch of our arm, *i.e.,* the seacoast service, would play a most decisive part. It should also be borne in mind that, whatever may be the future of the military organization of the country, *the defence of our harbors* must necessarily be foremost in the plan."

The chief distinction between the artillerist and the engineers was Whistler's implicit denial of the possibility of invasion. With the next war probably a naval one, he wanted to concentrate on securing naval bases and commercial harbors. There he was in accord with the engineers: "Referring to the question of time and efficiency, let it be remembered that to properly fortify our harbors would, under the most lavish expenditure, be a work of many years' duration, and that a reasonable expenditure for improvements from year to year would enable us to keep well abreast of the times. Concerning the question of expense, I submit, as a mere business proposition, that the people of the United States cannot afford to have the city of New York destroyed by a bombardment once in a century,—no, nor once in two centuries."

Arguments like Whistler's helped turn military thinking away from coastal defense, as promoted by the chief of engineers, and toward harbor defense. It was the way he wanted to spend appropriations, however, that showed the direction the artillerists were headed. Whistler proposed a future system of harbor defense that would include (1) earthworks "riveted with iron, together with iron turrets, or armored batteries"; (2) "exceedingly heavy guns"; (3) concentration of fire upon an approaching vessel; (4) torpedo systems, stationary and movable; (5) floating batteries or monitors stationed in the harbors; and (6) machines to throw dynamite or other explosives onto ships. Whistler's system would entail machinery for handling heavy guns; sighting towers to overcome heavy smoke; central fire control for volleys, electrically controlled; and electrical generators. For manpower, Whistler relied on the national guard.[5]

Whistler's thinking was still in the early stages of evolution, but it was oriented toward tools, unlike the Corps of Engineers' structural emphasis. The corps could not find a spokesman for its case when Whistler's article appeared. It let an infantryman, Lieutenant Arthur L. Wagner, speak for it. "Boston, Portland, Portsmouth, and Hampton Roads," he said, "are accessible to the largest men-of-war on the seas; and there is not a single important harbor on the Atlantic coast that does not possess a sufficient depth of water to pass heavily armored vessels with eighty-one-ton guns." Raiders, he claimed, could destroy $2 billion in property in a few days. Instead of the artillerist's fine-tuned assembly of technological improvements, Wagner offered structural measures. "In order to protect its harbors and leave its navy free to act on the offensive," he said, "the United States must defend its coast with fortifications. These works should be capable of resisting the most formidable naval attacks, should employ the most powerful ordnance, and serve as operating stations for submarine mines." Like the engineers, Wagner favored forts essentially as they were, elaborated with earthworks and turrets.[6]

The widening debate sparked public interest. President Rutherford B. Hayes raised the issue in his annual message for 1880, and Chester A. Arthur did the same in 1882. The result, finally, was a provision in the Fortifications Appropriations Act of 3 March 1885. It called for a board of experts to report what defenses were needed at what ports, and what kinds were "best adapted for each, with reference to armament, the utilization of torpedoes, mines, or other defensive appliances."[7] The board became known as the

Endicott Board after its president, Secretary of War William C. Endicott. The membership included the chief of ordnance, the chief of engineers, one officer from each of those services, two high-ranking navy officers, and two prominent citizens. The panel conducted the most searching study of American defense policies and systems ever made before 1900 and sent its results to Congress on 23 January 1886. It was a detailed product, saying what was to be done and what it would cost—over $126 million.[8]

To no one's surprise, the Endicott Board declared the nation's defenses to be in sad condition: "Without enlarging upon this subject, it suffices to state that the coast fortifications, which in 1860 were not surpassed by those of any country for efficiency, either for offense or defense, and were entirely competent to resist vessels of war of that period, have, since the introduction of rifled guns of heavy power and of armor plating in the navies of the world, become unable to cope with modern iron or steel clad ships of war; far less to prevent their passage into the ports destined for attack." The board continued, "It is of no advantage to conceal the fact that the ports along our seacoast . . . invite naval attack; nor that our richest ports, from their greater depth of water and capacity to admit the largest and most formidable armored ships, are of all the most defenseless."[9]

Condemning the nation's "supineness" in allowing its defenses to lapse, the board reviewed modern defense technology. That new hardware included floating batteries, armored-turret shore batteries, submarine mines and batteries to defend them, movable torpedoes and torpedo boats and support facilities, and heavy and rapid-fire guns. The board turned away from iron armor and toward steel. It then outlined what was proposed for twenty-seven harbors, arranged in priority. Hampton Roads ranked fifth on the list, after New York, San Francisco, Boston, and the "Lake ports."[10] The board proposed to begin with the first eleven cities on the list, addressing the secondary ports afterward.

At Hampton Roads, the plan called for masonry and earthworks, armor, structural metal, guns and mortars, carriages, submarine mines, and torpedo boats. The board based its cost estimates on the costs of iron but thought that "equivalent protection" in steel would "probably be the same." The meat of the Endicott Report was a set of eight appropriations proposals to procure gun steel, a federal gun factory, emplacement construction, armor and structural metal, gun carriages, floating batteries, submarine mines, and torpedo boats. The total first appropriation would be $21.5 million, to be followed by annual installments of $9 million until the work was completed. "[N]othing less will suffice even for a beginning," the board said. These figures shocked Congress, which did not even pass an appropriation for maintenance in 1886. It would be some time before the lawmakers decided to implement the Endicott Report.[11]

Congress's pause over the proposed system was fortunate. It was excessive, and not just because of cost. If all the works called for had been built, it would have taken 80,000 soldiers to man them fully—at a time when the army was stalled at around 25,000 enlisted men and Congress showed no interest in making it larger. After implementation began, the program was repeatedly reduced. The new weapons considerably outperformed predictions of the 1880s and also were more expensive than expected.

The guns were the most striking aspect of the Endicott period. Like their counterparts on the world's naval ships, the weapons had flat trajectories, with 15 degrees usually the maximum elevation. They eventually appeared in the form of 8-inch, 10-inch, and 12-inch breech-loading rifles. Many were mounted on the Endicott Plan's hallmark—"disappearing" carriages. The early models operated hydraulically, but that technique never was successful. Eventually the Ordnance Department settled on the Buffington-Crozier disappearing carriage developed by two of its officers, which operated by means of a counterweight sinking into a well beneath the gun. The weapons were lowered behind parapets for loading, then raised into position for firing, the recoil of which depressed them again. The parapets of the new works hid them, so the guns actually did disappear, at least from an enemy's point of view. Not as impressive visually but important in the early thinking were electrically operated mines and such weapons as dynamite guns, rifled 12-inch mortars, and so on. Many of them were passing fancies or outright failures and were dropped from the program in a few years.[12]

The weaponry reflected what was perhaps the fundamental spirit behind the Endicott Board's proposals—an almost boyish fascination with the new technologies, a romantic delight in new machines that was analogous to the Third System's earlier love of French military geometry. But the very absorption in current technology that made the board's plan so impressive also was its greatest flaw. The Third System had been designed according to principles thought to be timeless verities, but new ideas had made the verities old-fashioned. Similarly, the Endicott Board failed to take sufficient account of the prospect of change in technologies. The board proposed to make an enormous investment in the current state of the art without, evidently, considering what would happen if new advances made the current wisdom obsolete.

The mortars and dynamite guns were examples of what was about to happen. Large sums were spent on the former, while the latter never could be made practical. Both aimed at penetrating the upper decks of warships—mortars brutally with high-angle explosive shells, dynamite guns by tossing packages of the wondrous new "high explosive" via one mechanism or another. Ship decks of the 1880s were weak because the new guns had low trajectories, which made side armor more important than deck protection. By the time ships acquired armored decks, the mortars were obsolete and dynamite guns were stillborn. Ship designers armored their decks because gun designers developed high-angle weapons, but the Endicott Plan's shore defenses were designed to house and withstand low-angle gunnery. There was, accordingly, over the life of the Endicott period program a constant push and pull among competing technologies and tactics, a mutual evolution that outdated the whole system before it was completed. Yet the work went on, adjusted as well as could be, until the accumulated changes made a reconsideration necessary.

The board's fascination with gadgetry also blinded it to the political implications of what it proposed to do. No member of the board, it appears, considered whether there really was a threat of attack on American ports sufficient to justify such expensive defenses. The panel's collective assumption was that an attack was inevitable—or, perhaps more deeply implicit, that an attack would come unless the United States poured its treasury into defensive measures. To be sure, the European nations were beginning an arms race, especially in naval forces (and America soon joined in, if cautiously), but realistically that development reflected imperial and Continental competition, not designs against America.

Nevertheless, with the Endicott Plan the United States adopted a military policy of defense (or deterrence) which required very expensive measures to meet a threat that no one clearly identified. If the country was vulnerable, it was so only to the naval potential represented by the European arms race. Yet that arms race was evolving through its own action and reaction of technology, tactics, and politics. The naval potential of the unidentified enemy therefore would inevitably surpass the defensive potential of the Endicott program, locked into the technical thinking of the 1880s. As the foreign naval potential grew, the United States would appear increasingly defenseless, thus justifying further expenditures. The American military and naval establishments had found the keys to the treasury. It was just a matter of time before they learned how to work the lock.

The Endicott system also changed the role of the engineers in defense construction. Emphasis shifted from forts or emplacements to the weapons they housed, as the former became less and the latter more sophisticated and expensive. The powerful gunnery and low ballistic trajectories dashed any hope of continuing patterns established after the War of 1812. The new works were simple but massive, their crests at or just above ground level. Concrete frontal walls as much as 20 feet thick, behind 30 or more feet of earth or sand, made the installations almost invisible and invulnerable (for the moment) from seaward. The gunners and the ordnance experts now dominated American coast defense.

The deliberations of the Endicott Board caused a sea change in the Corps of Engineers. The fort builders, except for a few reactionaries, all but gave up on the Third System. They recognized that weapons technology had conquered fortification construction, and almost overnight their language changed. Paeans to the forts of the past gave way to increasingly sophisticated discussion of work that now must be shared with others. In 1887 the Board of Engineers commented on the fact that Congress had taken no action on the Endicott Report program:

> From the difference of opinion which exists among certain non-military experts as to the character of armor to be used in land defenses it has been argued that the whole subject of sea-coast defense is in an unsettled and tentative condition and that the policy of inaction now existing should still continue. But the facts will not warrant this conclusion, as more than nine-tenths of the armament recommended for our sea-

coasts is not to be mounted behind iron protections, but in rear of earthen covers surmounting and shielding the masonry magazines, bomb-proofs, and store-rooms. Particularly is this true of the rifled mortars . . . and there is no reason why the erection of the batteries required for them should be delayed a single month. Neither is armor required for guns mounted on lifts or disappearing carriages; in a word, proper sums may judiciously be expended and much progress toward placing our coasts in a defensive condition may be made, indeed must be made, before the question of armor demands consideration.[13]

The role of the engineers appeared destined for further reduction in 1887, if Professor Peter S. Michie of West Point had his way. He proposed a Heavy Artillery Corps modeled on the Corps of Engineers and would give it the entire job of building, maintaining, and manning harbor defenses. His idea saw partial realization in the establishment of the Coast Artillery Corps in 1907, but the engineers continued to handle most construction. Meanwhile, the War Department chafed at the absence of appropriations for its new enterprise. Congress provided for Watervliet Arsenal, the Endicott Plan's gun factory, in 1888, but that put nothing on the seacoast.[14]

Finally, an appropriations act of 18 August 1890 provided money for new works at New York, San Francisco, and Boston. The chief of engineers wanted additional funding to start construction at Washington and Hampton Roads as well. He also asked for money for land acquisition at all five harbors and reported that the engineers had begun working on mining casemates and cable galleries at the first three locations.[15]

Funding started off far lower than Endicott and his colleagues wanted, but at least it started. The first appropriations ended the large polemical literature on the nation's vulnerability and transformed the debate into one about relative technical roles in harbor defense. By the time that discussion was finished, the engineers had moved far over, not always happily, to make room for the artillerists and ordnance experts—and increasingly for the navy. One of the last professional articles to belabor the country's vulnerability to "sudden war" was by Brigadier General John Gibbon, an old artillerist, who published his piece in 1890. He had little to say about engineering and concluded: "The efficient means, therefore, of averting the dangers of a sudden war are a formidable navy and a perfected torpedo system for each harbor." He went only so far beyond these measures as to add revolving-turret guns. Admiral David Dixon Porter spoke in a more optimistic spirit, commending the new fortifications. Unlike in the past, he said, the forts now planned could defeat armored ships in a fight. Captain James Chester impressed his audience in 1892 with an up-to-date study of sophisticated technologies for position finding and detection—a great change from the point-and-shoot days of smoothbore muzzle-loaders.

Colonel Henry L. Abbott of the engineers offered a study of the construction program in 1894, then involving thirty seaports. He suggested that about seventy other places would justify "inexpensive earthworks" and that the Third System forts still might be useful with modern guns. More positively, he proclaimed the fear of invasions groundless and advised that site planning be guided by the threat of raids.

Engineer Peter C. Hains also spoke up that year, after it had been proposed that all coastal defense be consolidated in the navy; he wanted to keep the shore-based program in the army. More specifically, he wanted most of it in the Corps of Engineers, which should undertake mine laying, operation of torpedoes and torpedo boats, and, of course, the construction and maintenance of forts. All he would leave to the artillery would be the firing of guns. The artillerists, however, had the last word. Said Lieutenant George W. Van Deusen: "But when we speak of coast defenses, we do not mean fortifications alone, but all the accessories connected with them, such as submarine mines, moveable torpedoes and torpedo boats, and in some cases, armored coast-defense vessels or floating batteries."[16]

The engineers' role was still a major one. They soon embarked on the most expensive fortifications program they ever handled. By 1891 they had purchased sites and begun emplacements at New York, Boston, San Francisco, Hampton Roads, and Washington, while ordnance provided for government manufacture of high-powered steel guns and let contracts for others. By 1892 mining casemates at Washington, Hampton Roads, and San Francisco were completed, and four others were near completion. "The problem of adequate coast defense has practically been settled,"

Secretary of War Redfield Proctor said. "Fortunately it has not been a question of party politics."[17]

It may not have been a question of party politics, but it was one of budgetary economics. If the Endicott Plan had been followed as originally presented, nearly $98 million would have been spent on guns and emplacements by 1895. The actual total was only $10.6 million, averaging about $1.5 million a year. At that rate it would take another twenty-two years to finish the job. The War Department was not satisfied, but Congress was unwilling to spend more.[18]

Affairs improved with the appropriations act of 6 June 1896, which granted $2.5 million for gun and mortar batteries, "of which sum not exceeding one hundred thousand dollars may be expended for the construction of necessary buildings connected therewith." The latter provision was to provide decent barracks for garrisons still huddled in casemates, but by the end of the year the $100,000 had not been released to the Corps of Engineers. Instead, that sum and later housing appropriations, for coastal forts and inland posts alike, went to the Quartermaster Department, which provided the housing. The 1896 act also granted to the secretary of war, for the first time, authority to contract for work to be paid out of future appropriations. The chief of engineers was still not satisfied; he wanted more appropriations and contract authority both. Meanwhile, his engineers ventured into the gunners' technical territory when they developed a fire-control system for various installations and prepared to erect observation stations.[19]

Congress had stepped up the fortifications program because of souring relations with Spain, the buildup of German and Russian battle fleets, and a trend in naval construction that would culminate with the first all-big-gun, heavily armored battleship, HMS *Dreadnought,* in 1906. Funding increased further in 1897. The chief of engineers declared that he was losing control of his program because of congressional insistence that the work be done by contract. He wanted all jobs to be carried out by purchase and hire of materials and labor, to him the most economical course to follow. What he failed to say was that technical difficulties had appeared in the new works. The chief complained in 1898 that the artillery was not prepared for the sophisticated equipment turned over to it by the engineers, and he advised recruiting electricians for every battery. He revealed the start of a major dispute in 1899 when he alluded to difficulties over the design of range and position finders. Worse, seepage and dampness had occurred in the new works. All available preservation and maintenance funds went into attempts to waterproof the structures, at the expense of the remaining Third System forts. These preservation funds also were used to protect iron and steel from the ocean air and to care for the "large quantity of torpedo material accumulated as the result of the War with Spain."[20]

The seepage problem reflected a larger difficulty with the Endicott program. It was, simply, pursued faster than professionals could keep up with new technologies. Engineers in the United States were relatively new at concrete construction. Most of their background on the subject related to "natural" and "patent" cements, used mainly as canal liners and roofing. Portland cement concrete, however, was the preferred material for the new defenses. It was easily worked and formed and exceedingly hard and strong, but it generated heat when it set, especially in large masses, and it shrank while setting. Most important, it was not waterproof, both inherently and because of a tendency to crack during setting. The engineers, with increasing experience, would learn to contend with the seepage problem, but it would never go away.

The structures in the Endicott program were enormously thick masses of poured portland concrete, which aggravated the difficulties inherent in the material. Reinforcing techniques—the stabilization of concrete with rods or bars—had only begun to be explored, and it was not until after 1900 that the Quartermaster Department would build the first federal structures with reinforced concrete. If the technique had been available earlier, the engineers could have avoided many problems. Reinforced concrete was more resistant to shellfire, so the works could have been smaller in mass and therefore easier and cheaper to build. By the time that technique became known, the die had been cast, and along with it millions of dollars in construction money.[21]

The technical difficulties in the emplacements also reflected bureaucratic problems. The engineers, ordnance, and the artillery—three separate domains—declined to cooperate as well as they should have. The engineers had a long tradition of knowing what was best for the army—and especially an assumed monopoly on fortifications—so they caused most of the trouble. The four-pit mortar battery, four tubes to a pit, proved to be crowded and hazardous during firing of ordnance's proud new mortars, which anyway were not accurate enough for naval warfare. The engineers designed and built sighting and commanders' stations that the artillery rejected as unusable or dangerous. Early gun emplacements, especially, were designed without sufficient reference to ordnance's specifications for the guns and lacked enough room for the artillery to service the pieces. And finally, the engineers lost control of submarine mines after a raging struggle with the artillerists.

Although the pace of the work sometimes exceeded technical understanding or the level of bureaucratic harmony, it was still not as fast as the Endicott Board had wanted. Of 2,362 pieces of ordnance contemplated in the board's report, only 151 were in place at the outbreak of the war with Spain in April 1898, and there was a shortage of ammunition. Congress appropriated $50 million for "defense" before the war broke out, split between the army and the navy. The War Department spent almost all of its share on seacoast fortifications and armament, expecting the first events of the war to be off the American coast. The engineers had thirty projects approved and twenty-five underway by June 1898, along with many temporary batteries. Construction leaped forward, fueled by a series of appropriations during the year. By the end of the year the chief of engineers could point to ready emplacements for 288 heavy guns, 154 rapid-fire guns, and 312 mortars at seventy-one localities in twenty-nine harbors. As the war went on, his people overhauled old-fashioned armament and mounted it in several old forts, including Fort Monroe. In addition, they worked during the year to acquire, construct, and lay mine systems.

The wartime boom fueled the engineers' appetite. They reported at the end of 1899 that they had approved projects for thirty ports. They were considering additional localities in the United States, as well as the new island possessions, while work was underway at ten important harbors. An appropriation on 25 May 1900 added more money, bringing the total since 1890 to over $22 million for battery construction alone. Funds for maintenance, however, were low, and a serious maintenance backlog developed in many facilities.[22]

Appropriations subsided in 1901. Meanwhile, the engineers, artillerists, and ordnance experts engaged in furious quarrels over the range- and position-finding facilities developed by the engineers. The chief debate was over the siting of tall towers on low sites and the size of observation rooms. Work on the structures was halted by the secretary of war. Undaunted, the chief of engineers asked for an appropriation for them anyway, saying that the issues would be resolved. Preservation and maintenance problems in the emplacements worsened as that feud progressed, and another grew over unprotected battery commanders' stations. Congress increased appropriations again in 1902, and Inspector General Joseph C. Breckinridge, among others, was pleased, but he was concerned that not enough attention was being shown to maintenance or to defenses in the new island possessions. Congress tripled the maintenance appropriation in 1903, and the chief of engineers promised to put all the money into plugging leaks and correcting defects in the design of the new works. Then, as suddenly as it had arisen, the flow of money stopped after 21 April 1904. All the engineers could do was correct their errors of the previous fifteen years. Construction of additional works halted.[23]

President Theodore Roosevelt and Secretary of War William H. Taft decided that the Endicott program had gone far enough without reconsideration. Gunnery and ship construction had evolved faster than the assumptions underlying the program, and many of its conceptions were erroneous. Systems revealed to be misguided had been terminated but only only after expensive installations had been built; it had been possible to scale down the original plan but also too late in some cases. Before the government invested more in coastal defense, especially in the new territories overseas, the president and the secretary wanted to examine it in light of current technology. Meanwhile, the effort had produced impressive results, including new defenses of Hampton Roads.

Defense of Hampton Roads

The Endicott program did not itemize its work at Hampton Roads as taking place at Fort Monroe, although that was where most construction occurred. Fort Monroe retained its separate identity only for maintenance and special projects. Facing a new situation with new methods, the engineers termed Fort Monroe's additions "Defense of Hampton Roads." Changes in the engineers' thinking were presaged in 1880 when they lumped Forts Monroe and Wool together as "Defenses of Hampton Roads and

Gosport Navy-Yard." That plan, completed in 1884, upgraded the works but essentially kept them as they had been for decades; the most signal advance was reliance on breech-loading rifles (B.L.R.). The Board of Engineers' scheme for fifteen ports included for Hampton Roads:

Estimates for heavy guns and emplacements:

Three 2-gun turrets, at $600,000	$1,800,000
Six 100-ton breech-loading rifles, at $100,000 . . .	600,000
Ten iron casements [casemates], at $100,000 . . .	1,000,000
Ten 50-ton breech-loading rifles, at $50,000	500,000
Twenty 30-ton breech-loading barbette disappearing rifles, at $30,000	600,000
Twenty emplacements for 30-ton rifles, at $30,000	600,000
Sixteen 12-inch rifled mortars and emplacements, at $17,000	272,000
Total	$5,372,000

Estimates for emplacements, excluding armor, guns, and machinery:

Three turrets, exclusive of armor, at $220,000 . .$	660,000
Twenty emplacements for 30-ton, breech-loading barbette disappearing rifles, at $10,000	200,000
Sixteen emplacements for 12-inch rifled mortars, at $2,000	32,000
Total . . . .$	892,000

The chief of engineers briefly concentrated on the armored turrets in his list. He did not ask for additional construction money for fiscal 1886 but requested funds for turrets, including those for Hampton Roads.[24]

Such proposals faded when the Endicott Board went to work but did not disappear, as they influenced the Endicott Board's final program. Hampton Roads was fifth on the board's list of "ports [where] fortifications or other defenses are most urgently required." The Endicott Plan recommended floating batteries for two places—New Orleans and San Francisco. However, "the Board desires to point out that while not required at present, others may be useful to guard the eastern end of Long Island Sound and the approaches on that side to New York, and in Chesapeake Bay as an outer line of defence to Baltimore, Washington, and Hampton Roads." Floating batteries—a timeworn idea—were expensive, and even the board knew the country could not afford everything. Ultimately, none appeared at Hampton Roads or elsewhere. Nevertheless, the board did plan a lot for Hampton Roads:

> Turrets, armored casemates, barbette batteries, mortar batteries. Submarine mines will form a part of the defense. Eighteen torpedo-boats are recommended for the service here and in Chesapeake Bay.

| | Proposed Amount | | |
Caliber	Kind	Number	Remarks
16-inch	110-ton guns	4	B.L.R.
12-inch	50-ton guns	10	B.L.R.
10-inch	27-ton guns	20	B.L.R.
12-inch	Mortars	16	Rifled

The Endicott proposals reflected a belief that gunnery would not change in any fundamental way except in two respects—guns would become very heavy, and they would be predominantly breech-loading rifles. The ramifications of the latter trend were not yet fully appreciated. Designers were already at work on steel—not iron—weapons more sophisticated and powerful than anything known before. When they appeared, their dead weight was insignificant compared to their recoil, but the Endicott Board could do no better than describe Hampton Roads' defense by weights and measures, the only information on hand to translate into cost estimates. The 1886 estimates were broken down according to tons and cost of "Armor, wrought," tons and cost of "iron, wrought & cast, Structural metal," and cost of masonry and earth. Two "2-gun" turrets, an unspecified number of "10-inch casemates," twenty "disappearing & non-disappearing" guns, and sixteen mortars would adorn Hampton Roads. The location also would require 400 submarine mines, two "Operating Rooms &c.," and eight projectors in a system of "Electric lights, complete." Last, the board proposed three groups of torpedo boats, eighteen boats in all.[25]

The Endicott Plan was as out of scale for Hampton Roads as for any other place. Much of the scheme proved unnecessary when better information became available. Besides a general scaling down over the next twenty years, many specific pieces were omitted. Advances in ship design and naval tactics made mortar batteries obsolete by 1900, while dynamite guns never were practical. The most glaring deletion was turreted batteries. The United States never built them, except for one in the Philippines during the twentieth century.

On 2 August 1887 the Board of Engineers submitted estimates for fiscal 1889 for new works at the eleven most important cities. None of the items called for iron, because the board thought it could make faster progress on less expensive things. There were two typical plans for mortar batteries and disappearing-gun positions. For Hampton Roads the program would start with one 16-tube battery of 12-inch mortars and a disappearing two-gun 12-inch battery at Fort Monroe, along with one 12-inch disappearing gun at Fort Wool. On 15 June 1888 the board added a report on "Submarine defenses of Hampton Roads." The planning, if not the construction, was underway.[26]

The plans exceeded anyone's ability to carry them out, however. The few disappearing carriages in existence were experimental and did not work well. So the engineers continued to refine plans and figures. There were to be five classes of works mounting heavy ordnance, Chief of Engineers Casey announced in 1889: (1) mortar batteries, with and without scarp walls and flank defenses; (2) barbette batteries with guns on disappearing carriages; (3) barbette batteries with guns on vertical lift carriages; (4) iron-clad casemated batteries; and (5) iron or steel turrets. He then declared: "The efficiency and economy incident to the first three classes are so well determined that I am prepared to recommend their immediate construction at Boston, New York, Hampton Roads, San Francisco, and Washington, D.C., as the commencement of a comprehensive system of defense, which should be extended to other localities from year to year." The Board of Engineers, that same year, reported further plans for Hampton Roads and in 1890 advised purchase of sites there and at the other four cities.[27]

In one sense Casey was correct—construction could begin without resolving all questions on guns and carriages. On the other hand, Casey underestimated the extent to which more powerful and longer-range gunnery would dominate installation location and design. That may have become apparent to the engineers as early as 1890. When the money started to appear, they concentrated first on simple things and set aside facilities for disappearing carriages. Congress appropriated the first sums in August 1890 and February 1891. The first measure was directed to Boston, New York, and San Francisco; the second supported work at Hampton Roads. The lawmakers also provided for land acquisition. The engineers asked for condemnation proceedings at several places, including 44½ acres at Hampton Roads.

The first work was on shore facilities for electrically operated submarine mines, and on 24 November 1890 the Board of Engineers approved Peter C. Hains's project for a mining casemate and cable gallery at Fort Monroe. A year later it adopted his plans for emplacements for high-powered guns, although money was not yet available. The Endicott program had arrived at Hampton Roads and already had been scaled down. "The approved project of defense," the chief of engineers said of Hampton Roads in 1891, "contemplates, for the present, five 12-inch guns on lifts, ten 10-inch guns on disappearing carriages, thirty-two 12-inch mortars, and submarine mines operated from two mining casemates." Absent were such things as turrets, and mortars had doubled, but for the moment the plan for Hampton Roads was stable.[28]

By the end of June 1891 Hains completed concrete work for a mining casemate northeast of Fort Monroe and covered the squat little building with sand. Not so easy were emplacements for two 10-inch guns. Hains's aide Lieutenant Goerge A. Zinn gathered materials, laid a railroad track, built a bridge, erected storage bins for cement and aggregate, installed a concrete mixer and other physical plant, and ran a telephone line between the work places. Then the trouble began. Redoubt A, as it was first known—it later was renamed Battery Bomford, after George V. Bomford, longtime chief of ordnance—was in Fort Monroe's old redoubt, immediately north of the main work. It had been altered before work stopped in the mid-1870s and was a difficult case. Hains later related that in October 1891 "removal of the old redoubt with its concrete gun platforms and magazine was commenced and proved a difficult and expensive piece of work." It was not until March 1892 that mixing and laying of concrete could start. The demolition was such an ordeal that Zinn told the world about it in a magazine article later that year.[29]

Zinn finished the mining casemate in fiscal 1893, and Hains reported that "a storehouse for mining material is also provided." The concrete parapet for Redoubt A was finished and about half of the earthwork on its front. Construction of a third 10-inch-gun emplacement was authorized in December 1892, and Zinn built another railroad and concrete plant and began mixing and laying. The third 10-inch-gun emplacement—Redoubt B, later part of Battery Church—and completion of the first two-gun work ran into a problem when the army could not decide on carriages for the guns. Ordnance selected the Buffington-Crozier disappearing carriage in 1895, and work resumed. The three emplacements were completed in 1896, and their armament arrived soon after. The only things needed to have two of the 10-inch guns ready were ammunition cranes and some sodding on the earthwork face in front of the concrete structure.

The scale of the work was reflected in the supervisor's summary report of 1896: "The sand foundation for these platforms was thoroughly rammed and settled with water. The surface of the platforms was covered with 4,700 square feet of granolithic finish, 2 inches in thickness, in the composition of which 88 barrels of Portland cement, 20 cubic yards of fine broken stone, and 15 cubic yards of sand were used." That did not include the far more massive parapet structures built previously. Each platform itself consumed 664 yards of concrete, twelve hinge stones, six sills, six lintels, twenty steps (cut and set), six doors (made and hung), seventy-six steel I-beams (placed), thirty-two 2-inch steel bolts and fifty-six 1½-inch steel bolts, seventy-two straight iron plates, 216 feet of curved iron plates (to serve as beam rests), forty-two removable iron rods for railing around the loading platform, and iron ladders. The slopes around the platform took 1,577 cubic yards of sand, 20 of clay, and 20 of soil. The guns and platforms were turned over to the artillery in 1897. The two-gun work mounted Model 1888 M-II 10-inch rifles and the single-gun emplacement a Model 1888 M-I, all on disappearing carriages and capable of firing out to 12,259 yards.[30]

In March 1895 money was allotted for a battery of sixteen 12-inch mortars; a project was submitted in October. The office of the chief of engineers reduced its cost by placing the four four-tube pits in a single line north of the mining casemate, instead of in the usual cluster of pits in a quadrangle, an arrangement which also had been found to concentrate concussion to a dangerous degree during firing. Work began with clearing of trees and brush and construction of a concrete plant, three hoisting engines, repair shops, storage bins, and storehouses, and excavating sand for the foundations. The concrete work started during fiscal 1897, and the job was finished in 1898 after the engineers installed electrical lighting in the mortar battery and the gun emplacements.

The mortars were one of the curiosities of the Endicott program. The pits were just that—great square concrete holes open at the top and rear. The mortars were mounted four to a pit, with four and later two pits comprising a battery. The South Battery at Fort Monroe later was named for Robert Anderson, commander of Fort Sumter in 1861, and the adjoining North Battery for Adjutant General George D. Ruggles. The original armament was Model 1890 M-I mortars on twelve Model 1896 and four Model 1891 carriages (the latter changed to Model 1896 types in 1901). The rifled tubes had ranges between 2,210 and 15,000 yards, but with doubtful accuracy. The purpose was to bring plunging fire down onto warships, but changing ship design and tactics soon made that pointless. Hampton Roads never saw its second set of sixteen mortars.[31]

The gunnery appearing at Hampton Roads was of a type the Corps of Engineers had never dealt with before. Younger officers like Zinn and Hains's successor Thomas L. Casey, Jr., did well enough, but some of their superiors had difficulty. Not appreciating the power of the weapons they installed, they still thought in terms of the massiveness and abundant firepower that had guided the Third System. Hampton Roads would be heavily fortified, Colonel Henry L. Abbott of the Board of Engineers said in 1894, "thus directly covering the Navy-yard at Norfolk and the city of Richmond, and giving our fleet (supposed to be smaller than that of the enemy) a secure position where it may issue at will to operate on his lines of communication if he attempts to enter the bay. Such a disposition will compel him to mask our naval force with a superior fleet before he can attempt interior operations." The chief of engineers overlooked modern firepower and modern design alike when he described the defenses of Hampton Roads in 1896. "These works comprise two works of the older type," he said, "with outlying batteries. . . . Both are needed in the revised plans for defense."[32]

That philosophy of give up nothing and get more of the same guided further development at Hampton Roads. The original plan had been scaled down once, but in the late 1890s new works appeared as the opportunity presented itself. No general scheme was apparent other than the urge to add facilities, to spend whatever money was pouring in. In 1896 Casey submitted plans and estimates for a tower and shelter for a Lewis range finder. A contractor built the 41½-foot-high concrete-and-brick tower, and Casey turned it over to Fort Monroe's commander in April 1897. The structure was one of the developments that sparked the feud among the engineers, artillerists, and ordnance men, and it turned out to be a waste of effort. The tower vibrated at every shock, making it ineffective.

Casey dropped range-finder work in 1897 to build another emplacement for a 10-inch gun. Before long work was going on at every hand. The "National Defense" appropriation of March 1898 allowed for more 10-inch disappearing guns, a number of 4.72-inch rapid-fire guns, and one 8-inch breech-loading rifle on a non-disappearing carriage, the last "being of a temporary character." By July 1898 the engineers had completed platforms for two 10-inch guns, the 8-inch rifle, and three rapid-fire guns and had "three rapid-fire guns mounted & ready for service." A second mining casemate, inside the old fort, also was started and completed. Casey installed a new electrical power plant, wired all facilities, fired the mortars to test the platforms, and moved an 8-inch breech-loading rifle to the main fort. Because there was a war on, his staff and contractors obstructed the harbor entrance channel with 108 submarine mines, issued and enforced special navigation rules, and patrolled the minefields round the clock, making repairs as necessary.

The war ended in December 1898, but work continued the next year on a single 10-inch position in the bastion of old Fort Monroe and the two-gun battery of 10-inch rifles. Both were near completion by July 1899, as was the battery for four 15-pounder rapid-fire guns, and the garrison had mounted an additional 10-inch weapon. Work began on what would be Battery DeRussy, originally designed to house three 12-inch disappearing guns and one 10-inch gun on a disappearing carriage (the platform for the latter was not actually built). By that time the engineers had added three 10-inch guns and one 4.72-inch rapid-fire gun to the place's armament, along with the mounting of a fourth 8-inch rifle and the dismounting and removal of two others.

Workmen also set up a 36-inch searchlight on the parapet of Fort Monroe as part of the torpedo defense. In August and September 1899 they took up the mines and moved them to Fort Wool for storage. The mines were swept up with tarred rope, and all electrical equipment was dismantled, cleaned, and stored. Five mines were exploded as an experiment, and the engineers planned to disarm the rest. They anchored one down at Fort Wool and unscrewed the compound plug with a socket wrench attached to a drum rotated by rope from a distance. The fumes that greeted them when the plug fell out threatened danger, so the entire investment in mines was blown up off Fort Wool.

The amazing flurry of activity begun just before the war with Spain and continuing in its aftermath added much to Fort Monroe's arsenal. It included one old 8-inch rifle on a barbette carriage at the northwest corner of the old fort; one 10-inch emplacement to become the second half of Battery Church (for Albert E. Church, professor at West Point), completed in 1899; two 10-inch emplacements in Redoubt C, eventually Battery Eustis (for Colonel Abraham Eustis, first head of the Artillery School), completed in 1899; four 4.72-inch rapid-fire emplacements on the barbette tier of front four over the east gate, completed in 1898 and 1899, called Battery Gatewood (Lieutenant Charles B. Gatewood of the Geronimo campaign), and soon replaced with Model 1898 British Armstrong rapid-fire guns; four 8-inch rifles temporarily mounted on the fort rampart; three 12-inch disappearing guns in a battery between Redoubts B and C, finished and 1901 and later called Battery DeRussy (Colonel René E. DeRussy, engineer at Hampton Roads before the Civil War); and one 10-inch emplacement (not named) in the old bastion near the east gate, with an experimental "depressing gun" which was soon taken down.[33]

The engineers returned to Fort Wool in 1899. It and the older parts of Fort Monroe were repaired that year, and the engineers planned emplacements for two 12-inch guns on the Rip Raps. More work first had to be done at Fort Monroe, where in 1900 and 1901 laborers started Battery Irwin (Lieutenant Douglas S. Irwin, killed in the Mexican War), completed in 1902 and 1903, mounting four 15-pounder rapid-fire guns bearing on the channel opposite Fort Wool; Battery Parrott (Captain Robert P. Parrott of the Ordnance Department), begun next to Battery Irwin in 1900, completed in 1905 and 1906, and mounting two 12-inch disappearing guns; and Battery Montgomery (Major Lemuel P. Montgomery, killed at Horseshoe Bend in 1814), between Batteries Church and DeRussy, mounting two 6-inch Model 1900 rifles on barbette carriages protected by armor shields, completed in 1904.

The last two projects were especially interesting. Battery Parrott's construction required the destruction of the old water battery. The new work sported the most powerful weapons ever mounted at Fort Monroe. The 59-ton guns (range 17,000 yards) became thundering showpieces of the Coast Artillery School established at Fort Monroe in 1907. As for Battery Montgomery, its 6-inch guns (range 13,077 yards) were the last of the Endicott period weapons to remain in service in the United States, not removed until 1948.[34]

While the engineers were in their last flurry of construction, they faced and lost further challenges to their role in harbor defense. Legislation reorganizing the army on 2 February 1901 transferred torpedo defense to the artillery. The engineers at Hampton Roads gathered the stuff together, and it was in the artillery's hands by the end of 1901. The engineers made their last contribution to gun pointing in 1904, when they built a number of datum points—creosoted piles with horizontal reference strips five feet above mean low water, varying in width according to distance from sighting towers. They would continue to build sighting and range-finding facilities, but others would specify what those should be.

They retained maintenance of the works. That was a monumental challenge by 1904. Battery Anderson, one of the mortar emplacements, leaked prodigiously. Captain E. Eveleth Winslow made a special study of the problem, then built a corrugated iron ceiling inside the rooms. It leaked also, and Winslow advised other engineers against that remedy. In 1905 he reported a new construction method to prevent condensation and divert leakage. Involving hollow brick, air spaces, and vapor barriers, it was simple and effective. The chief of engineers published Winslow's report for other designers, and the term "air space" became part of the fortification lexicon. Unfortunately, the improvement arrived too late to do much good, because most works had been built. It was useful in later construction in the tropics, however.[35]

The last major Hampton Roads project during the Endicott period was at Fort Wool. Planning had proceeded to build a battery for 12-inch guns. Tests in 1902 persuaded the engineers that they could never build a foundation there to withstand the recoil of such weapons. The plans therefore were revised to provide two 6-inch rapid-fire guns on disappearing carriages. Unfortunately, no such carriages for that weapon existed, so work was deferred again. Work did begin in 1902 on reconstruction of the wharf and on a concrete emplacement for four 3-inch guns (15-pounders) on the Norfolk side. The project involved the destruction of all casemates of Fort Wool except those at the extreme west end. The trace of the new works, however, followed that of the castle on the Rip Raps. The four-gun Battery Henry Lee (for Light-Horse Harry Lee) was completed in 1905. Work began in January 1903 on Battery Jacob Hindman (officer of the War of 1812), mounting two 3-inch Model 1902 rapid-fire guns facing the channel, completed in 1908. The adoption of the Model 1903 disappearing carriage and Model 1903 6-inch gun permitted work on three two-gun batteries (Frederick Claiborne, Alexander Dyer, and Horatio Gates), completed along with Battery Lee.

Fort Wool did not have much of a history thereafter. Three artillery companies were stationed there in 1905 during army-navy exercises, then removed. One company arrived in April 1917 to manage one end of an antisubmarine net stretching to Fort Monroe. That garrison left at the end of World War I, taking with it the 6-inch guns. Except for a caretaker, Fort Wool sat alone until World War II.[36]

The engineers—and their colleagues in the artillery and ordnance branches—left some of their most impressive achievements in the Endicott installations at Fort Monroe. Their work was durable, and much of it survives today. It would have measured up to the naval gunnery of its time.[37] That circumstance was about to end, however.

The Taft Report

The Endicott system was 50 percent complete by 1900, with twenty-five harbors adjudged ready to face attack. Where there were garrisons—most installations were in caretaker status pending a war—the gunners did occasional target practice. Disappearing rifles bobbed up and down in a thundering dance, throwing smoke and fire and noise over the parapets, then dropping back into their nests. Ponderous mortars belched fountains of smoke and flame and spat missiles high into the air. It was all very impressive, but was it effective defense?

Not everyone thought so, and increasingly the officers of the navy developed ideas of their own. Their arm had long been promoted as the first line of defense. Under the influence of Alfred Thayer Mahan of the Naval War College (son of Dennis Hart Mahan of West Point), a doctrine of "command of the seas" evolved, leading to a new strategy. Coastal defense by the army became more narrowly defined as the protection of the navy's bases, rather than all major ports and harbors. The navy began to play a major role in the mechanics of harbor defense and in the years before 1940 became the dominant decision maker regarding what points required protection by the army.[38]

Sorry events during the war with Spain made it imperative that there be more consultation between the services. Inspector General Breckinridge went so far as to question the wisdom of keeping the two departments separate. "Perhaps it was a wise provision of our institutions" and worked well enough, but he pointed out one instance after another of failures, near disasters, and simple lack of cooperation. So he suggested forming a board of army and navy officers to establish joint procedures and standards for interservice cooperation. The Joint Army and Navy Board was instituted on 17 July 1903, to include four men from each service. When combined with the navalist sentiments of President Theodore Roosevelt, the establishment of the board meant that the seamen would increase their role in defense.[39]

Readjustment of the balance between army and navy was a long-term matter. In the short run technological progress and the budget had overtaken the Endicott program. It had been reduced incrementally since its inception, and at the beginning of the new century more fundamental changes were in the offing. As Chief of Engineers George L. Gillespie explained in 1902: "Nearly fifteen years have elapsed since the adopted scheme of coast defense was formulated by the Endicott Board. At that time the rapid-fire gun was in its infancy, and ships were characterized by extremely heavy guns and great thickness of armor. With the development of the rapid-fire gun and the increase in the resisting powers of armor a material change has taken place in ship construction, necessitating corresponding changes in the details of coast defense."

Some of those changes had preceded technology, for budgetary reasons. Gillespie—ignoring the proposals of his predecessors—boasted that the United States had avoided "costly experiments" with armored turrets. The gun designers had produced a general reduction in calibers and an increase in shell penetration and explosiveness. Accordingly, many plans were revised to reduce the number of heavy guns and mortars in favor or more rapid-fire weapons.[40] Their high velocity, penetration, and rate of fire were such that they could concentrate more destruction on an enemy vessel—in particular, the smaller ships and boats thought to threaten mine defenses—than more expensive big guns.

Big-bore guns were not the only ones to suffer cutbacks. Legislation on 15 May 1900 and subsequent measures forbade building any more rifled-mortar batteries. The engineers objected, believing that the weapons were accurate enough for their purpose, close-in defense. Nevertheless, they had become nearly useless against the latest naval technology, with its stronger upper decks and longer-ranged guns. As for an oddity less successful than mortars, Gillespie announced in 1902 that "during the fiscal year the Secretary of War directed the sale of the obsolete dynamite guns, and no further reports upon dynamite batteries will be submitted." By 1904 the chief of engineers could point to continued progress on rapid-fire ordnance and declare it "now a matter of first importance." It produced greater results for less investment.[41]

The sweeping changes in technology, evolution in naval doctrine, and pressing need to start work in the overseas territories caused the president to stop asking for fortifications appropriations in 1904. On 31 January 1905 he organized the Taft Board, named after its president, Secretary of War William Howard Taft. The board transmitted its report on 1 February 1906, after which work resumed.

There was not much Taft and his board members could do about the Endicott system, because most of the work had been completed. Their greatest influence came in construction overseas, marked by two developments. One was the wider spacing of the batteries and the adoption of a new 14-inch gun, larger but more workable than its 12-inch predecessors. The other was the revival of the seacoast mortar after 1911 but in smaller numbers—two tubes per pit, four per battery, nearly all installed overseas. Within the United States the Taft Report led to a reduction in the number of pits per mortar battery and later in the number of mortars per pit. About 1906 most of the original four-pit batteries with four tubes per pit were divided into eight-tube batteries of two pits each. A few years later the number of mortars was reduced from four to two per pit, and half the weapons were scrapped—except at the Fort Monroe pits, which continued to be used for training.

The Taft Report emphasized searchlights, communications, powered ammunition handling, and other accessories recommended by the Endicott Board but neglected thereafter. Most important was the adoption of aiming systems using optical instrumentation, target data processing, and electrical transmission of sighting and gun-pointing information. The best system gave each battery two widely spaced "base end stations," or sighting locations, which offered simultaneous sighting and calculations of angles for transmittal to a central battery computing room. The information supplied gave present and predicted target locations and when corrected produced aiming directions to be transmitted to each gun emplacement or mortar pit. The new aiming system gave shore guns superiority over those afloat until about World War I. It also represented one more surrender of control over defense design on the part of the engineers. The new systems were devised by experts in ordnance and the artillery. The engineers found themselves confined to building the towers and other facilities that others specifed.

There was one other important outfall of the Taft Report. That was the addition of the defense of the Chesapeake Bay entrance to the program. For the first time, technology made closing that gap conceivable, if not immediately possible.[42]

Large appropriations for batteries did not follow immediately on the issuance of the Taft Report; even funds for maintenance were low. More money did appear for searchlights, ammunition hoists, and electrical gear. Finally, on 27 May 1908 the engineers received money for gun and mortar batteries—the first such funds in four years. Modernization of older emplacements, fire-control facilities, electrical and searchlight installations, and the rest continued to receive priority over batteries. Construction appropriations fell victim to an economy drive during a period of national economic depression, and no more appeared until March 1911. By 1912 the chief of engineers was upset that he had received only $305,064 in the years since the Taft Board had added nearly $11 million to his program. His protest earned him some funds in February 1913, but Congress did not appropriate money for the Chesapeake Bay entrance. Finally, Congress added funds for new construction in June 1914 and other money to modernize existing facilities. Before implementation of the engineers' part of the Taft program could get underway, however, it was overtaken by World War I.[43]

Entrance to Chesapeake Bay

The entrance to Chesapeake Bay was not neglected in the Endicott program, which included a mortar battery on Willoughby Point for the Hampton Roads area. In 1891 the Justice Department condemned and seized forty-seven acres and two rights-of-way there and transferred the property to the Corps of Engineers.[44] The Willoughby Point battery was intended to guard the southern side of the entrance to Chesapeake Bay. It soon fell out of the program, however, and was forgotten.

In 1900 the Corps of Engineers' only new proposal was a preliminary plan to defend the bay entrance at Cape Henry; the secretary of war approved the concept for a project in 1901. That was the end of the venture until 1906, when the Taft Report advised adding the entrance to Chesapeake Bay to the list of points to be defended. It was the last major addition to the program within the continental United States.[45]

Nothing happened to it for some time, however. In 1909 the chief of engineers asked for money to complete structures for mines at all places except Galveston, Texas, and the Chesapeake Bay entrance, "where, owing to the present status of the defenses and garrisoned posts, it is not considered advisable to build the torpedo structures immediately." The secretary of war cut the torpedo item and several others out of the budget altogether.

Cape Henry was tended by the winds until 11 December 1912, when Chief of Staff Leonard Wood told the House Appropriations Subcommittee on Fortifications: "On the Atlantic coast the most important piece of fortification work to be done is the proposed defense of the entrance of Chesapeake Bay. Next in order would be the completion of the fire control and searchlights for the fortifications already constructed." The Chesapeake Bay entrance, in other words, was in the army's view the only important hole in the defensive system. The current plan for Cape Henry called for land acquisition and installations for eight 12-inch mortars, four 14-inch rifles, and four 6-inch rifles. Present fortifications, Wood averred, were not adequate, because they could not protect the bay entrance or nearby cities from invasion. The new proposal for Cape Henry would be sufficient to fight off the latest seagoing technology and to give the navy a secure base at Hampton Roads.[46]

The War Department asked for money during fiscal 1914 to buy about 300 acres at Cape Henry, with construction to follow. The proposal set off much discussion among the congressmen about defense technology, but there was no opposition to a complete big-gun installation at the location. Brigadier General Erasmus M. Weaver, chief of the Coast Artillery Corps—at whose instigation the engineers' plans for 12-inch guns had been upgraded to 14-inch models—offered the only suggested alteration. He wanted the weaponry upgraded further to newer 16-inch cannons.

Congress discovered that it had already appropriated some money for land acquisition at Cape Henry in 1909 (nobody else had noticed either), so on 13 February 1913 it added the balance and told the War Department to buy the land. Negotiations with the landowners were unsuccessful because their prices were too high, and the haggling ended in condemnation, with an award to the owners higher than the funds available. Congress provided the extra money on 17 June 1914, and the land passed into federal ownership.[47]

The engineers had the property but could not build on it until funds became available; these appeared late in 1916. Not much progress was made before the United States entered the war in Europe in April 1917. Part of the property eventually was covered with temporary facilities later known as Fort Story. As for heavy-duty coastal defense, the engineers had other things to do. Two coast artillery companies moved to Cape Henry early in 1917, followed by four 6-inch guns on field carriages.[48]

World War I marked the end of the Endicott and Taft periods in coastal fortification. It rather abruptly terminated Corps of Engineers construction around Hampton Roads and Chesapeake Bay—except for rivers-and-harbors projects, which continued at a high level through the war.

The Coast Artillery Corps instituted temporary expedients. In March 1917 the coast artillery commandeered property on Fishermans Island, near Cape Charles opposite Cape Henry, and moved in two companies with four 5-inch guns. Hampton Roads bustled with military activity during the war and was the largest artillery training center on this side of the Atlantic. The gunners laid an antisubmarine net between Fort Monroe and Fort Wool in March 1917 and another near Thimble Shoal in August; they operated both until December 1918. Most reflective of a coming change in defensive technology was the opening of an antiaircraft training and development program at Fort Monroe, along with installation of antiaircraft guns at Fort Monroe, Newport News, and Hopewell, Virginia. Military ground forces now faced a threat from the sky; if it persisted, it would change coast defense forever.[49]

The Future of Coast Defense

The chief of the Coast Artillery Corps announced in 1914 that most approved defense projects were complete. Since 1888 the United States had spent $143.7 million for a comprehensive system, with its only serious weaknesses being at the Chesapeake Bay entrance, the defense of Los Angeles, and some batteries not yet in place at Manila Bay, Hawaii, and the Canal Zone.[50]

Whether all the expenditure served any real purpose—more important, whether the system met the needs of the present— remained to be demonstrated. All involved believed that the effort had been worthwhile, but the engineers alone thought that it had not gone far enough. By 1915 the engineer chief's annual pleas for budgets echoes those of his predecessors in the 1870s and 1880s. The war in Europe lent credence to old arguments about the country's vulnerability to foreign raiders. In 1916 the watchword of the day became "Preparedness." A supplemental appropriations act of 6 July 1916 revived funding for battery construction and land acquisition. Chief of Engineers William M. Black justified these projects by saying that the country must guard against landings of forces that could attack its harbor defenses from behind. He wanted to supplement the fixed fortifications with obstacles and field entrenchments and a force of mobile railroad artillery—another new development being demonstrated in France.[51]

His thinking had no room for the distractions posed to potential invaders by the ground war in Europe or for the complications in commerce raiding and naval warfare then being introduced by torpedo-firing submarines.

The approach of American entry into the war in 1917 brought more money for battery construction and separate funds for "land defenses." The situation then called mainly for temporary measures and finally for transfer of most resources to the American Expeditionary Force embarking for France, where the Europeans were so mired in their great bloodbath that they had no prospect of attacking the United States. Appropriations ceased after June 1917.

After the armistice, the chief of engineers declared that there were no lessons from the war that would justify fundamental changes. The older Endicott works, he said, should be modernized, but they remained the keystone of the country's defense, supplemented by railroad guns. Almost as an afterthought, he said that shielding the works from aerial observation and the addition of antiaircraft defenses might be advisable. Small appropriations appeared in March 1919, but the program fell victim to postwar economizing; all unobligated funds reverted to the Treasury on 30 June 1920. It would be impossible to develop emplacements for the newest heavy, long-range guns, which were now the theoretical order of the day.[52]

Frustration over minuscule budgets drove the engineers once again into the past. The world war, said Chief of Engineers Lansing H. Beach in 1920, demonstrated that the principles set forth by the Board of Engineers for Fortifications in 1826 still held true: "These principles, enunciated almost 100 years ago, have been proven by the war to be as sound to-day as when originally stated." So firm was he in holding to old ways that he denounced railroad guns, dismissed military aviation altogether, and ignored submarines, against which shore defenses were useless. The chief improvement needed, he said, was the adoption of guns of 16-inch caliber or larger—once again, what had gone before, more and bigger.[53]

Perhaps what Beach really wanted was to reclaim the dominance in coastal defense that had slipped away to other branches of the military and naval establishment. That loss was aggravated by further insults from new weapons technology, and especially from flying machines. The investment of the current generation of engineers in the thirty-year round of defense building was almost as great as that of their predecessors in the Third System. They would not let go of their creation lightly.

The coastal works of the Endicott and Taft periods were as vulnerable to attack from the air as their predecessors had been to rifled artillery. Moreover, their vulnerability predated the advent of aerial bombardment. Improving naval technology—carriages and recoil mechanisms in particular—so greatly increased the range and power of seaborne batteries by World War I that the coastal works were open to high-angle plunging fire. In response, by the end of the war the Coast Artillery Corps had divided its functions into fixed harbor defense, mobile seacoast artillery, and antiaircraft gunnery.[54]

The artillerymen knew that fortifications once again must give way to a new generation of men and machines. The engineers failed to share the gunners' vision.

ⅬⅬⅬⅬⅬⅬⅬⅬⅬ **7** ⅬⅬⅬⅬⅬⅬⅬⅬⅬ

The End of Fortifications

Everyone has his day and some days last
longer than others.
—Winston Churchill, 1952

Army construction, except for fortifications and most
work in the war zone, was during World War I the province of the
War Department Construction Division, an expansion of the construction arm of the Quartermaster Corps. In the war's aftermath
a move to consolidate the Construction Division into the Corps of
Engineers, and thereby have but one construction organization in
the army, failed. So did an attempt to create an independent
Construction Corps.

The National Defense Act of 1920 returned construction to the
quartermasters, except for "fortifications." These remained the
province of the engineers, but there was little engineering to do on
them for some time. The definition of theory and practice in coastal
defense passed to the Coast Artillery Corps, the navy, and the gun
designers. Moreover, the military establishment underwent two
decades of budgetary decline. There was no revival of general
construction activity until after 1926, when the quartermasters
embarked on a new housing program. It terminated for fiscal
reasons in 1933.[1]

A New Age of Coastal Warfare
Under the National Defense Act the Corps of Engineers
generally was confined to rivers and harbors. Fortifications were
the responsibility of the army's geographical corps area commanders, who had engineers on their staffs. In practice, however, civil
works districts of the Corps of Engineers provided engineering
services for the corps areas. The Norfolk district engineer became
the "Engineer in Local Charge of the Harbor Defenses of Chesa-

peake Bay" in October 1932. The assignment arrived none too soon, because two hurricanes in August and September 1933 seriously damaged Fort Monroe. A total of $1.6 million in emergency-relief money was made available for projects there. The engineers and quartermasters built new wharves and other facilities and restored the terrain with landfills and protective measures. The major part of the appropriation went for a "million-dollar seawall" built to prevent further storm damage. Construction began in March 1934 and was completed in December.[2]

The restoration of Fort Monroe was made possible by federal efforts to combat unemployment during the Great Depression. While that work was going on, the quartermasters also received relief money. One of their "million-dollar" projects for the construction of bases for the Army Air Corps began in 1932 at Langley Field, north of Hampton Roads.[3]

Langely Field was well known to the public because just after the war it was the seat of General William Mitchell and his aviators in the Army Air Corps. They were fierce advocates of strategic air power and at their most extreme proclaimed that future wars would be waged entirely in the air, overtopping a country's defenses to strike at cities and factories.

Tactically, the airmen claimed that forts and guns had no real value in coastal defense because aircraft could engage and defeat an enemy fleet far offshore, out of the range of land-based cannons. Mitchell and his crews demonstrated aviation's potential by bombing and sinking captured German and obsolete American warships off Hampton Roads.

Hampton Roads, in fact, was a hotbed of arguments over the future role of military aviation. The war had scarcely ended before navy fliers took off and landed on ships there, foretelling naval aviation. The coast defense problem, then, would not be the weight of enemy gunnery but the ability of hostile naval fliers to bypass shore guns and devastate cities and harbors. The airmen held that defense against attack from the skies was essentially impossible. The Coast Artillery Corps disagreed and assumed leadership in antiaircraft gunnery and theory at its school at Fort Monroe. It began with batteries to protect conventional coast defense emplacements and expanded to the point where antiaircraft gunnery became the coast artillery's chief function. The role of the engineers in defense planning declined further. Nevertheless, in 1942 the corps found itself busy building antiaircraft gun platforms at various locations around Hampton Roads.[4]

Paralleling the development of antiaircraft principles was that of the mobile seacoast artillery, also called railroad artillery. This movable defense system was something that postwar chiefs of engineers believed held great promise. Its implementation, however, rested with the Coast Artillery Corps, which had used railroad guns in France during World War I. This activity also focused on Hampton Roads, where Fort Eustis, next to Newport News, became the national headquarters of the railroad artillery in the early 1920s.

The coast artillery at first drew on World War I leftovers, and in the 1920s various forms of mobile artillery played an increasing role in the gunners' ideas of how the coasts might be defended. Fixed positions became steadily more vulnerable to attack from the air. Movable guns would be harder for a naval fleet to attack with aircraft simply because their exact position could not be determined in advance and they were free of the structural evidences that made fixed positions easy to locate. Railroad and tractor-drawn guns proliferated, and some weapons were even removed from fixed installations. Of railroad types, the 8-inch gun, 12-inch mortar, and a new-model 14-inch gun predominated. Tractor-drawn weapons, however, eventually became the coast artillery's chief form of mobile defense against ships. The most important was the Model 1918 155-millimeter gun, which remained the chief weapon to the end of World War II.

The movable weapons were not as accurate as their fixed counterparts. That drawback, however, was partly solved by the engineers, who built simple circular-track fixed platforms for them at likely locations, including all around Hampton Roads, available for use when needed. The mobile artillery, as experience demonstrated, was not in the final analysis mobile enough to eliminate fixed emplacements altogether. Most of the movable guns went into storage late in the 1920s and early in the 1930s, although some 8-inch railroad guns remained in service at the Chesapeake Bay entrance. By 1940 mobile coast artillery was a dead letter, except for the 155-millimeter gun. When defense money again became available, the movable guns were mostly scrapped in favor of fixed emplacements, and the engineers resumed a larger role in defensive construction.[5]

One adjunct of fixed coastal defenses was submarine mines. The engineers were removed from that work early in the twentieth century, and by 1930 it was solidly a coast artillery enterprise. The Submarine Mine Depot moved from Fort Totten, New York, to Fort Monroe in 1930, following the Submarine Mine School, which migrated there in 1908. By 1932 theory and training in mining operations were completely overhauled. During 1942 the coast artillery, coast guard, and navy installed and operated mines and antisubmarine nets in the Hampton Roads area. The Corps of Engineers, however, provided base facilities.[6]

The Last Days of Fixed Defenses

As had happened after the Civil War, technological developments after World War I led to a new generation of fixed defenses. Once again gunnery dictated structural design, rather then the reverse. Two developments were of particular importance. One was the Model 1917 high-angle barbette carriage that increased the range of Endicott-period 12-inch guns. The last new batteries—comparatively simple concrete structures—for those carriages were completed by the late 1920s. The other innovation was a completely new 16-inch gun, which became available during the 1920s. That great cannon could throw a one-ton shell over thirty miles and was arguably the most powerful gun in the world.

The adoption of the 16-inch gun was complicated by the international disarmament movement of the early 1920s. Battleships then played the same role in public fears about armaments that nuclear-tipped missiles assumed after the 1960s. Their threat was believed to have been reduced by the Washington Naval Armaments Treaty of 1922, which scrapped many warships in the world's navies. That treaty reduced the motivation for a new round of coast defenses, but it also contributed to their development, for a number of naval 16-inch guns, only slightly less powerful than those of the army, became available for coast defense. By the late 1920s they were the standard weapon for large, fixed coast defenses and remained so until the end of World War II.

The heavy, accurate, and high-angle gunnery of the day outmoded the high-parapet, open gun emplacements developed from

the 1880s to World War I. Never again would such a great investment in expensive and massive works be advisable. The emplacements of the 1920s were simpler and more standardized than any
before, and dispersion became their chief design characteristic.
The gun stations were also well concealed, although not from the
air. In fact, their circular 360-degree concrete traverses resembled
bull's-eyes when seen from above. The gunners, however, aimed at
ships and trusted the antiaircraft service to protect them from
planes.[7]

The simplicity of the installations allowed work to proceed fairly
fast and economically. Congress appropriated funds for battery
construction on 3 March 1921 and 30 June 1922, and the chief of
engineers predicted the completion of all work in fiscal 1923. New
big-gun batteries did not appear in great numbers around the
entrance to Chesapeake Bay, however. As late as 1922 it was
regarded as infeasible to close that 12-mile gap in the coastline.
The Corps of Engineers built Batteries Pennington and Walke,
simple concrete platforms for four 16-inch howitzers—not to be
confused with 16-inch guns—at Fort Story (Cape Henry) from
1920 to 1922, but there was no money for anything more. The
batteries at Cape Henry had a range of 24,540 yards, while current theories called for keeping enemy ships 35,000 yards from
Cape Charles and Cape Henry to close the bay entrance.

For many years the area's chief armament remained at Fort
Monroe, outdated and out of range of the bay entrance. It was not
an effective defense of Chesapeake Bay, but it was enough for a
grand demonstration in 1924, the centennial year of the artillery
school at Fort Monroe. Training exercises had been held since the
first batteries were installed, but those of 1924 were especially
impressive and were viewed by a host of dignitaries. Ships and
aircraft passed by towing targets, and all the guns of Hampton
Roads opened up from Forts Monroe, Eustis, and Story. They were
supplemented by railroad and other mobile artillery, submarine
mine demonstrations, and antiaircraft cannon.[8]

Construction of 16-inch batteries stopped after 1922 because
money ceased arriving and because the War Department paused
to reconsider coastal defense in an age of aircraft and modern
navies. The Chesapeake Bay entrance still lacked big 16-inch
emplacements in 1928, the best thing available being the 16-inch

howitzers at Fort Story. The Coast Artillery Corps was sensitive to charges that the bay remained vulnerable. Large public exercises that year presented the message that, thanks to modern gunnery, never again would it be possible for an enemy to invade the bay as had happened in the War of 1812. That was the last service exercise for the 16-inch howitzers until 1941. All training exercises for the coast artillery at Hampton Roads and vicinity came to an end with the cancellation of the 1933 exercises, for reasons of budget and the loss of officers to the Civilian Conservation Corps.[9]

The halt in fixed coastal defenses was only temporary nationally, although it endured for years around Hampton Roads. In 1923 the War Department issued a study which established defense policy until World War II. Either a larger navy or more aircraft could provide better protection than fixed harbor defenses, the document said, but those forms of defense would not be economical. "When it comes to preventing enemy ships from sailing into a harbor and taking possession," the study concluded, "the cheapest and most reliable defense appears to be guns and submarine mines."[10] The likelihood of an enemy attack was not a subject of the report. The Endicott program had established its own momentum in national military thinking, and the threat was assumed to exist. To meet it, the United States would continue to do basically what it had always done—fortify the harbors.

Drastic reductions in the budget, especially during the depression, did not allow the engineers to do much more than study emplacement design for several years. Nor, except for the circular tracks for 155-millimeter guns, was very much required in new construction until the late 1930s. Much of the coast artillery inventory went into storage or caretaker status, and the country put aside thoughts of war in a spirit of isolationism. Several major harbors, according to current theories, remained inadequately guarded, including the entrance to Chesapeake Bay. Another such place was San Francisco, where in 1937 the engineers started to build a new type of emplacement for 16-inch guns, designed to provide cover from air attack. The first example was a pair of 16-inch guns in great casemates 600 feet apart, connected by galleries housing magazines and support facilities. To withstand bombs or projectiles falling from above, the roofs were of reinforced concrete 8 to 10 feet thick, covered by 20 feet of earth. The cover was increased directly over the casemates, from which only the barrels of the cannons protruded, further shielded by concrete eyebrows. Virtually all heavy emplacements built afterward followed that pattern.[11]

San Francisco was the only place thus guarded for some time. Budgets for a new round of defenses were not forthcoming, even after the start of limited mobilization in 1938. The chief of the Coast Artillery Corps grew increasingly concerned after war erupted in Europe and in May 1940 warned the chief of staff: "With but few exceptions our seacoast batteries are outmoded and today are woefully inadequate. Nearly every battery is outranged by guns aboard ship that are of the same caliber. More alarming than this is the fact that every battery on the Atlantic Coast, and all but two of the batteries on the Pacific Coast, have no overhead cover so are open to attack from the air."[12]

The fall of France that spring galvanized the War Department into action. In June 1940 the Harbor Defense Board recommended the 16-inch and new 6-inch guns as standard for harbor defense and proposed twenty-seven new two-gun casemated batteries for 16-inch guns. The Chesapeake Bay entrance was supposed to receive ten batteries of 12-inch and 16-inch guns, mounted either in concrete casemates or in bunkers with armor shielding. The fear of aircraft now dominated installation design. The engineers over the following years installed about 100 new batteries around the country. The guns typically were 500 feet apart, sheltered by concrete and earthen canopies.

The weapons were governed by hundreds of observation and fire-control structures, many of them towers, extending for miles along the coast from each battery. Those built around Hampton Roads were simple enough towers or housings on elevated points. On more isolated stretches of coast elsewhere, and especially in New England, the engineers outdid themselves with imaginative camouflage. Many control facilities were masked to resemble such things as silos or farmhouses.

Although the program showed imagination in its details, its general conception was decidedly old-fashioned. It was little more than the Endicott Plan or even the Third System—witness the rebirth of casemated emplacements—dressed up in new materials. At a time when the potential enemies and allies had already been identified, and the general strategy of taking the war overseas had been determined, there seems to have been a hope of impressing the public that the War Department was on the job, defending the potential home front. When the war actually developed, however, its real needs drained resources from coastal defense, the need and utility of which seemed steadily less pressing. Simply put, the time had arrived when it was obvious that aircraft, ships, and flexible ground forces could serve the harbor defense need better than fixed emplacements, and without the permanent investments the latter entailed.

As the engineers played out their part, the coast artillery also overhauled its program. Virtually all pre–World War I antiship weaponry was discarded in favor of four guns in the 1940s. They included the 16-inch gun (mostly the naval model), a new 6-inch gun with a 15-mile range, the high-angle 12-inch gun developed in World War I, and a few 8-inch guns, taken out of storage and put on new carriages. Antiaircraft gunnery underwent a transition to mostly a new 90-millimeter model developed early in the war. The reforms simplified supply, maintenance, and training, while the Allies received many discarded weapons.[13]

The new defense programs made themselves felt around the entrance to Chesapeake Bay. Since 1917 the bay's harbor defenses had been Fort Monroe and Fort Story. In September 1940 the army established Fort Winslow, on Cape Charles and Fishermans Island. It was renamed Fort Custis in answer to local sentiment in February 1942 and Fort John Custis the following October to avoid confusion with the reactivated Fort Eustis, reviving after closing early in the depression. The most important development was implementation of the Harbor Defense Board's June 1940 program for Hampton Roads. That plan called for two 16-inch batteries and two 6-inch batteries at Fort Story, two 16-inch batteries and two 6-inch batteries at Fort Winslow (Custis), one 16-inch battery at Fort Monroe, and one 6-inch battery at Fort Wool. However, the work got off to a slow start, as it did nationally, owing to competing priorities and the evolution of defense thinking.

The 16-inch howitzers at Fort Story returned to action on 10 June 1941. A crowd of 500 spectators and newsmen looked on, distracted from the continuing absence of 16-inch guns at the Chesapeake Bay entrance. The engineers had not begun work when, late in the summer, the War Department limited construction to those coastal batteries that could be completed by 30 June 1944, because of competing mobilization demands. In the absence of new work by the engineers, the coast artillery relied on mobile artillery. A two-gun battery of 8-inch railroad guns and a four-gun 155-millimeter battery were emplaced at Fort Story before the end of 1941. Another 8-inch railroad battery and the 155-millimeter batteries at Fort Monroe were earmarked for Fort Winslow, where the engineers built trackage and emplacements. Last, a radar set was erected at Fort Monroe, on Battery Montgomery.[14]

The Japanese attack on Pearl Harbor—the airplane's supreme insult to conventional harbor defense thinking—carried the United States into the war in December 1941, and attention to defenses around Hampton Roads redoubled. The coast artillery laid its first line of mines between 8 and 16 December and moved the 8-inch railroad battery from Fort Monroe to Fort Winslow by the twenty-ninth. Antiaircraft defenses also were activated on 8 December, and one company of troops moved out to Fort Wool, where the engineers erected temporary wooden buildings for them in February. The big fortifications, however, were slower to arrive, and many did not appear at all. Competition with other war needs held up weapons production for the new batteries, and that put a brake on construction.

By September 1942 it was obvious that the program would not be completed, and perhaps should not be. The army, with the concurrence of the navy, scrapped ten of the 16-inch projects already deferred. During the next two years similar cutbacks squashed the 6-inch program. In November 1942 nine more low-priority 16-inch batteries, on which construction had not started, were deleted from the War Department project list. Among them were the battery at Fort Monroe and one at Fort John Custis. The coast artillery, meanwhile, replaced the old 8-inch guns at Fort Custis with new railroad guns in August 1942.[15]

Hampton Roads and vicinity not only failed to receive new armament but lost much of what was on hand to wartime scrap drives. In 1942 Fort Monroe alone supplied 2 million pounds of scrap, including the mortars of Batteries Anderson and Ruggles. In 1943 the 12-inch rifles of Battery Parrott were donated, followed in 1944 by the 12-inch rifles of Battery DeRussy. Even trophy German guns from World War I went onto the scrap heap. Howitzers captured from the British at the Battle of Saratoga in 1777 would have followed them but for the National Park Service, which wanted them for the Saratoga National Historical Park.

The chief exception to the general reduction in weaponry was the introduction in 1943 of 90-millimeter antiaircraft guns, adapted for use against torpedo boats. One battery, radar-controlled, replaced Battery Parrott's 12-inch rifles, while two batteries appeared at Fort Story and a fourth on Fishermans Island. By the end of the war the 90-millimeter guns were Hampton Roads' predominant antiship gunnery.[16]

The local engineers office, based at old Fort Norfolk, finally started Hampton Roads' new fixed defenses in 1943, beginning with three permanent concrete positions for .50-caliber machine guns on the parapet of Fort Wool. Two 37-millimeter guns also were installed there, replaced in 1944 by 40-millimeter guns. Two old 6-inch guns were removed from Fort Wool in April 1943, when work started on a long-range barbette 6-inch battery (the new model) on the site of Battery Gates. That was completed by the end of the war, but no guns were mounted there. Fort Wool was in caretaker status from 1944 to 1947, while two of its four 3-inch guns went to Fort John Custis; the others were moved to Battery Irwin, Fort Monroe, in 1946.[17]

Other emplacements were harder to come by. In August 1943 the War Department stopped work on thirteen 6-inch batteries, including those at Fort Wool and Fort John Custis, because of greater needs for gun barrels in the field artillery. One Fort Wool emplacement was mostly complete, the other had hardly started. By 1944 the progress of the war nearly put an end to the heavy emplacement program, but not before the engineers and contractors began work on San Francisco-type casemates and galleries at Cape Henry and Cape Charles. Battery Ketchum, at Fort Story, was completed in the summer of 1945 and on 27 November fired twenty rounds to test the 16-inch weapons and the works. Battery Winslow, the 16-inch battery at Fort John Custis, also was completed and proof fired late in the year. But the war was over. As the great guns thundered, Army Ground Forces began a review of harbor defense which would abandon a century and a half of tradition founded in 1794.[18]

Dismantling the coast defenses began right after the war. The two remaining 3-inch guns of Battery Lee on Fort Wool were removed to Battery Irwin, Fort Monroe, 31 May 1946, but not for defensive purposes. They were wanted for salute firing during the transfer of the Army Ground Forces headquarters to Fort Monroe. They were still in place forty years later, the only Coast Artillery Corps armament remaining at Fort Monroe.

After Army Ground Forces completed its review of the subject, Fort Monroe lost its role in harbor defense; most of its guns and all mines were removed in November 1947. Battery Gates, for 6-inch guns on Fort Wool, had never been armed. Its retention was considered for a while, but soon it too was discontinued.

The entire system of coastal defense, which had as much as called the Corps of Engineers into existence in the early nineteenth century and occupied much of its energy thereafter, was fairly well abandoned in 1948. World War II demonstrated that large invasions were possible without seizure of port facilities and that naval armament would be more effectively used in support of amphibious operations than in shelling of cities and harbors; coastal defenses had not prevented the Allies from invading Hitler's "Fortress Europe." There was, in addition, the complicating factor of the submarine, the war's chief commerce raider, virtually invisible to on-shore gunnery. Aviation also scrambled the calculations of the War Department; aircraft and antiaircraft measures had been both more effective and less decisive than their respective promoters had promised. By 1950 the last harbor defense command was out of existence, followed shortly by the Coast Artillery Corps itself.

The weapons of Hampton Roads already were on their way to the dump. The 6-inch guns of Battery Montgomery were scrapped in March 1948, and the battery itself was demolished for housing construction in the early 1950s. The 90-millimeter guns at Battery Parrott were designated a saluting battery in 1948, then removed in 1950. Only the 3-inch guns of Battery Irwin remained, but saluting ammunition for them soon was unavailable; field howitzers assumed the ceremonial chores. The army engineers, who had created them, delivered the last blows to their fortifications at Fort Monroe during the 1950s. In March 1951 Company A, 981st Engineers Battalion (Construction), removed Batteries Bomford and Barber. Another such unit demolished Battery Eustis in March 1959. Battery Humphrey and the original redan of old Fort Monroe soon followed. A long tradition had come to an end, but Fort Monroe itself remained. For many years it was the headquarters of the Continental Army Command, then of the Training and Doctrine Command. In the aftermath of the Vietnam War, it was the only active United States Army fort surrounded by a moat.[19]

War and Transformation

Mobilization and war had an enormous impact on the Hampton Roads area, proportionately larger than on any other metropolitan region in the United States. The changes attracted a federally funded sociological investigation, which summarized:

> From the settlement of Jamestown in 1607 to the present, war has been a potent factor in shaping the lives of people living along the shores of Hampton Roads. . . . The impact upon the region and its people, however, of the First and Second World Wars, though fought in distant lands and waters, was probably even sharper than the earlier campaigns. The building and repair of ships; the training, embarkation and debarkation of soldiers and sailors; and other war activities absorbed, especially in World War II, most of the existing human and other productive resources of the region, brought in thousands of migrants from other parts of the country, and necessitated the construction of vast military and industrial plants, public works, and housing.[20]

Mobilization and war also transformed the Corps of Engineers. The first impact was the transfer of Army Air Forces construction in the United States and overseas (except Panama) from the Quartermaster Corps to the Corps of Engineers, ordered on 20 November 1940. The engineers also acquired the airport construction program of the Civil Aviation Administration. Bigger changes happened on 16 December 1941, after legislation transferred the Construction Division of the Quartermaster Corps into the Corps of Engineers. "While requiring extensive reorganization," said the chief of engineers, "the combining of the construction activities formerly carried out under the two offices has been effected with minimum dislocation, and the basic functional plans for the efficient prosecution of civil works have been maintained intact." The process was not that simple, of course, because the two groups had contrasting organizations and operational methods, although both worked on construction. It was also painful for officers who had to remove the quartermaster wagon wheels they had worn throughout their careers and replace them with engineers castles. In any event, most personnel of the two bodies were civilians, and the demands of the national emergency overrode personal sentiments.

The changes in the Corps of Engineers were dramatic. Following on the long drought in funding, the absorption of the gigantic general construction program of the quartermasters diluted the engineers' fascinations with forts to insignificance. More wrenching as a whole, however, was the suppression of the corps's interest in river-and-harbor improvements and its reconversion into a military organization. Over the past century, during the rise and fall of its fortifications programs, the Corps of Engineers had become an agency mostly civilian in staffing and oriented mostly toward such civil works as the dredging of waterways; forts were its only real ties to the military establishment.

There was a strong feeling in the Corps of Engineers that civil works should go forward in World War II, as they had in World War I. Brigadier General Brehon B. Somervell—a former engineer, current head of the quartermaster construction program, and future chief of Army Service Forces—had other ideas as consolidation approached. Priorities, he said, must be clearly understood: "The construction work of the Quartermaster Corps overshadows overwhelmingly the construction work being done by the Corps of Engineers, and military construction both in amount and importance bids fair to continue to be the major effort of the Engineers for several years. . . . Under no circumstances should the less important, slow moving, civil works be permitted to dominate the reorganization for vital, fast-moving and extensive requirements."

That philosophy was not shared in the Corps of Engineers until the bombs fell on Pearl Harbor. The districts tried to keep the civil works going, but by the middle of 1942 civil projects fell off dramatically, and until their revival in 1945 the corps was substantially a military-construction agency, dominated by the construction outlook of the quartermasters. The scale of the new organization was described by Chief of Engineers Eugene Reybold at the time of the consolidation:

> Consolidation of the construction functions of the Quartermaster Corps and the Corps of Engineers brings together organizations that are engaged in a 3½ billion dollar defense

program, embracing projects in every State, in Alaska, Panama, and Hawaii, and at island bases throughout the Western Hemisphere. This vast program engages the attention of some 600,000 individuals, including contractors' employees. If we were organized as a corporation we should be the world's largest. In fact, this merging of functions involves about the same number of persons as might be affected if the United States Steel Corporation should decide to combine with the Bell Telephone Company.[21]

The corps mostly reorganized itself according to the existing construction program of the quartermasters and found itself in a whole new sphere of work. The chief of engineers placed geographical division engineers in charge of the whole program and realigned the divisions to match the zones of the quartermaster construction program, which were coterminous with the army corps areas. New districts were created, boundaries adjusted to match the work load, and the districts became the effective operating arm for all projects. Subdistricts called areas proliferated for direct supervision of construction works, thereby conforming to the organization established by the quartermasters and in turn grafting it onto the countrywide establishment of the engineers. Rather than simply reorganizing, the Corps of Engineers of old was subsumed into a vastly different and larger organization with a stupendous variety of things to do.

By 1943, after most major projects were completed, activities around Hampton Roads remained widely varied:

Norfolk and Vicinity: Air Force facilities; Eastern Defense Command, Norfolk District; Harbor Defenses of Chesapeake Bay; Nansemond Ordnance Depot; "Passive Protection."

Newport News and Vicinity: Camp Patrick Henry; Fort Eustis; Hampton Roads Port of Embarkation; Harbor Defense of Chesapeake Bay, Newport News vicinity; New Point Comfort; USO [United Services Organization], Norfolk.

Camp Pickett: Camp Pickett.

Richmond and Vicinity: Camp Lee; Richmond Deepwater Terminal; Lumber Yard Operations; Richmond Air Base; Richmond General Hospital; Richmond Quartermaster Depot.

Langley Field: Air Force Facilities; Langley Field, AWS

[Aircraft Warning Service] and VHF [Very High Frequency] Control System.[22]

The consolidated construction organization became the army's landlord and housekeeper as well as its builder. The two functions overlapped at Fort Monroe, where engineers officers maintained buildings, grounds, and utilities throughout the war, and others erected building after building. One of the most important of the latter was the Submarine Mine Depot, built by the engineers in 1940 for the special unit of the same name that was in charge of the development, procurement, and supply of coast artillery mining equipment, cable, and floating plant. The depot also manufactured electrical material.[23]

Another important facility was the Little Creek Mine Base. Mine planters and the mine batteries of the Chesapeake Bay defenses used it throughout the war as a base for field operations, although facilities for the inner mine field remained at Fort Monroe. Construction work started in 1941 and concluded in 1942, and it was characteristic of the pace of mobilization (and Corps of Engineers tradition) that its first occupants had no barracks.[24]

The construction program during World War II, except for the few gun emplacements, was unlike anything the engineers had done before. Instead of forts and dams and water-control structures, now they built barracks, houses, utility structures, hospitals, warehouses, and factories. The largest enterprise of all was the Hampton Roads Port of Embarkation, established at Newport News in June 1942, with the new Camp Patrick Henry its major staging area. The total capacity of the facility's housing was nearly a quarter-million men, together with all equipment for units moving overseas. It was almost an entire seaport community built from the ground up, from wharves and cranes to the latrines.[25]

Besides building construction, Corps of Engineers experts spent over two years in pioneering research at Langley Field on pavements for runways to receive "very heavy" bombers—specifically the B-29 then on the drawing boards.[26] The war ended with the dropping from B-29s of two atomic bombs, designed and assembled by an arm of the Corps of Engineers, the Manhattan District. That dramatic event underscored what had been happening to the national defense program since before the Civil War: Fortifications gave way increasingly before men and machines, and at last fixed defenses must be abandoned as the keystone of national security. In their place stood the more complicated processes of military construction and procurement.

The Corps of Engineers would never be the same. It had a permanent peacetime military works program to match its civil works program, and its activity would no longer be characterized by fits and starts on fixed coastal defenses. Those were gone and in their place stood a comprehensive design and construction program. The Corps of Engineers had become the largest general construction organization in the world.

L'Envoi

The Corps of Engineers lost touch with its roots in fortification during and after World War II, and not just because the old fascination was diluted by general military construction. Forts had become curious relics of a bygone age. Aircraft called for new responses to the defense problem, and the logical successor to coast defense of old was the network of radar stations the engineers built for the air force. They were followed for a while by the Nike antiaircraft missile system, for which the engineers built an installation at Hampton Roads in the mid-1950s. Long-range rockets with nuclear warheads soon made that obsolete, and although the corps was heavily involved in developing antiballistic missile (ABM) facilities in the 1960s and 1970s, they also were made futile by technology and politics. During the 1980s the old coast defense idea gained revival in President Ronald Reagan's "Strategic Defense Initiative" for antimissile defenses orbiting in space. Whether "Star Wars" (as it was called) would become a Fourth System was problematic, but one thing was clear—the Corps of Engineers would not dominate it.

More generally, the corps left its forts behind because—despite its wider role in military construction—it reverted to civil works, building dams and improving rivers and harbors. That waterway development, curiously, was how the engineers made their most important contribution to Hampton Roads' status as the greatest concentration of military and naval assets in America and one of the country's outstanding ports.[27]

Hampton Roads and vicinity were never harmed by an enemy after 1813. Whether that fact was owing to the defenses the engineers built there is debatable, as there was no real foreign military threat. Perhaps it was fortunate that the works were never tried, because in each period those defenses reflected the tendency of military experts to prepare for the last war in the delusion that they were making ready for the next. In vastly improving the naturally fine waterways of the Hampton Roads area, however, the Corps of Engineers facilitated the growth of the naval facilities there as well as the civil commerce and in the end made the region and the country stronger with or without forts.

The Corps of Engineers fortifications programs may have faded, but they left a strong legacy in American military history. Throughout the nineteenth century, when the army was small and mostly scattered in the wilderness chasing Indians, the Corps of Engineers remained its strongest connection to its real purpose— national defense against modern enemies. The enemies and the threat they posed may have been imaginary; but from its pulpit at West Point and in its emphasis on outward defense, the corps kept the army from sinking entirely into the military barbarism that its Indian fighting tended to encourage.

The several phases of fort building established other traditions as well. One was the tendency of large military programs to become larger, more expensive, seemingly endless, and ultimately self-serving—not to mention obsolete before completion. Fortifications in the nineteenth century and early twentieth century may have achieved that condition as compensation for an otherwise general American penchant for keeping the military organization and budget as small as possible. Whatever the cause, the tendency was more apparent in military programs during the decades after World War II, when the penchant was reversed.

More enduringly, the Corps of Engineers proved, beginning in the Third System, that it was possible to win financing for its military schemes by offering extravagant warnings about the dangers the nation faced—and equally extravagant claims about the military's ability to avert those dangers if it had its budgetary way. Success breeds success, so it should be no cause for wonder that into the 1980s the American military staked its claims to the treasury by retailing the evil intent of purported national enemies and the horrible character of their weapons—and by vouchsafing the ability of the American military to avert enemies and their armaments both if it received the money it sought. The ultimate defense would forestall the ultimate weapon.

For all the generations of predictions, however, no enemy fleet raided American ports after 1814. Nevertheless, thanks to the example of the fort-building Corps of Engineers, the United States has based its military programs for nearly two centuries on preparing for threats that never materialized—indeed, may never have existed. The habit continues in the present age, with this generation of soldiers and civilians—like those before—fearing its own dangers but wondering at the folly of its fearful ancestors.

Abbreviations

AA	*American Anthropologist*
AHR	*American Historical Review*
AR	Annual Report
ARBE	Annual Report of the Board of Engineers
ARCE	Annual Report of the Chief Engineer
ARComGen	Annual Report of the Commanding General
ARIG	Annual Report of the Inspector General
AROrd	Annual Report of the Ordnance Bureau
ARQMG	Annual Report of the Quartermaster General
ARSecWar	Annual Report of the Secretary of War
ARSurGen	Annual Report of the Surgeon General
ASP-MA	*American State Papers,* Class V, *Military Affairs*
BAE	*Bulletin of the Bureau of American Ethnology*
BE	Board of Engineers
CAJ	*Coast Artillery Journal*
CVSP	*Calendar of Virginia State Papers*
CWH	*Civil War History*
DA	Department of the Army
Heitman	Francis B. Heitman, *Historical Register and Dictionary of the United States Army, from Its Organization, September 29, 1789, to March 2, 1903,* 2 vols. (1903; rept. Urbana: University of Illinois Press, 1965)
JER	*Journal of the Early Republic*
JFH	*Journal of Forest History*
JMSIUS	*Journal of the Military Service Institution of the United States*
JUSA	*Journal of the United States Artillery*
Lewis	Emanuel Raymond Lewis, *Seacoast Fortifications of the United States: An Introductory History* (1970; rept. Annapolis: Leeward, 1979)
MA	*Military Affairs*
MAD	Military Affairs Document

ME	*Military Engineer*
MVHR	*Mississippi Valley Historical Review*
ND	Norfolk District (Office)
OCE	Office of the Chief of Engineers
PAAS	*Proceedings of the American Antiquarian Society*
PJCAMP	*Periodical Journal of the Council on Abandoned Military Posts*
PMHB	*Pennsylvania Magazine of History and Biography*
RBE	Records of the Board of Engineers
RO	Report(s) of Operations
ROCE	Records of the Office of the Chief of Engineers, Record Group 77, National Archives
USNIP	*United States Naval Institute Proceedings*
USWD	United States War Department
VC	*Virginia Cavalcade*
VHR	*Virginia Historical Register*
Virginia Guide	United States Writers' Program, Virginia, *Virginia: A Guide to the Old Dominion*, American Guide Series (1940; rept. New York: Oxford University Press, 1947)
VMHB	*Virginia Magazine of History and Biography*
Weigley	Russell F. Weigley, *History of the United States Army* (New York: Macmillan, 1967)
Weinert and Arthur	Richard P. Weinert and Robert Arthur, *Defender of the Chesapeake: The Story of Fort Monroe* (Annapolis: Leeward, 1978)
WMQ	*William and Mary Quarterly*

Notes

1

1. Weinert and Arthur, 2–17; Wesley Frank Craven, *The Southern Colonies in the Seventeenth Century, 1607–1689* (Baton Rouge: Louisiana State University Press, 1949), 65–66; *Viriginia Guide,* 32–34.

2. See *Virginia Guide,* 9–12; Nevin M. Fenneman, *Physiography of the Eastern United States* (New York: McGraw-Hill, 1938).

3. Maurice A. Mook, "The Anthropological Position of the Indian Tribes of Tidewater Virginia," *WMQ,* 2d ser., 23 (1943): 27–40; James Mooney, "The Siouan Tribes of the East," *BAE* 22 (1895); James Mooney, "The Powhatan Confederacy, Past and Present," *AA,* n.s., 9 (1907): 129–52; C. C. Willoughby, "The Virginia Indians in the Seventeenth Century," ibid., 57–86; J. R. Swanton, "Early History of the Eastern Siouan Tribes," in *Essays in Anthropology Presented to A. L. Kroeber* (Berkeley: University of California Press, 1936), 371–81. See also Harold L. Driver, *Indians of North America,* 2d ed. (Chicago: University of Chicago Press, 1969), 80, 106, 159, 302, 324, 355–56, 479–82.

4. Gary B. Nash, "The Image of the Indian in the Southern Colonial Mind," *WMQ,* 3d ser., 29 (1972): 197–230; Herbert I. Priestly, *The Coming of the White Man, 1492–1848* (New York: Macmillan, 1929); *Virginia Guide,* 23–30; Robert F. Berkofer, Jr., *The White Man's Indian: Images of the American Indian from Columbus to the Present* (New York: Knopf, 1978), 14–15, 18–21, 118, 129–30, 134; J. P. Kinney, *A Continent Lost—A Civilization Won: Indian Land Tenure in America* (Baltimore: Johns Hopkins University Press, 1937), 12, 14, 21, 28, 38, 40; Craven, *Southern Colonies,* 75–83.

5. Louis Morton, "The Origins of American Military Policy," *MA* 22 (1958): 75–82; Douglas Edward Leach, *Arms for Empire: A Military History of the British Colonies in North America, 1607–1763* (New York: Macmillan, 1973); Harold L. Peterson, *Arms and Armor in Colonial America, 1526–1783* (New York: Bramhall House, 1956); Lyon G. Tyler, ed., *Narratives of Early Virginia, 1622–1675* (Baltimore: Johns Hopkins University Press, 1908), 119–204; "Military Census of Virginia," *VMHB* 7 (1900): 364–67; "Acts, Orders, and Resolutions of the General Assembly of Virginia, July 1, 1644," ibid., 23 (1915); Oliver L. Spaulding, Jr., "The Military Studies of George Washington," *AHR* 29 (July 1924): 675–80; Bernard Knollenberg, *George Washington: The Virginia Period, 1732–1775* (Durham: Duke University Press, 1964); James Thomas Flexner, *George Washington: The Forge of Experience, 1732–1775* (Boston: Little, Brown, 1965); Douglas Southall Freeman, *George Washington,* 7 vols. (New York: Scribner, 1948–57); Ivor Noël-Hume, *Martin's Hundred* (New York: Knopf, 1982). See also Frederick Stokes Aldrich, "Organization and Administration of the Militia System of Colonial Virginia" (Ph.D. diss., American University, 1964).

6. James L. Stokesbury, *Navy and Empire* (New York: Morrow, 1983), 13–49; Craven, *Southern Colonies;* Max Savelle and Robert Middlekauff, *History of Colo-*

nial America, rev. ed. (New York: Holt, Rinehart and Winston, 1966); Charles M. Andrews, *The Colonial Period of American History,* 4 vols. (New Haven: Yale University Press, 1934–38), vol. 1; Thomas J. Wertenbaker, *Virginia under the Stuarts, 1607–1688* (Princeton, N.J.: Princeton University Press, 1914); Morton, "Origins of American Military Policy"; Leach, *Arms for Empire;* Irene A. Wright, "Spanish Policy toward Virginia, 1606–1612," *AHR* 25 (1919–20): 448–79.

7.   Craven, *Southern Colonies,* 27–92; Stokesbury, *Navy and Empire,* 40–42; *Virginia Guide,* 32–75; Matthew Page Andrews, *The Soul of a Nation: The Founding of Virginia and the Projection of New England* (New York: Scribner, 1943); Edmund S. Morgan, "The First American Boom: Virginia, 1618 to 1630," *WMQ,* 3d ser., 28 (1971): 169–98; Edward K. Chatterton, *Captain John Smith* (London: John Lane, 1927); John Smith, *The General History of Virginia, New England, and the Summer Isles* (London: Michael Sparkes, 1624).

8.   Weinert and Arthur, 3–6; Bruce Grant, *American Forts Yesterday and Today* (New York: Dutton, 1965), 60; Kinney, *Continent Lost,* 28, 38, 40; Marion L. Starkey, *The First Plantation: A History of Hampton and Elizabeth City County, Virginia, 1607–1887* (n.p., 1936), 10; Wright, "Spanish Policy." See also Robert Arthur, *History of Fort Monroe* (Fort Monroe: Coast Artillery School, 1930).

9.   Weinert and Arthur, 8–13.

10.   Ibid., 14–17; Grant, *American Forts,* 60; Lewis, 15; "Fort George," *VHR* 1 (1848): 20–23.

11.   Thomas J. Wertenbaker, *Norfolk: Historic Southern Port* (Durham: Duke University Press, 1931), 61–63.

12.   Ibid.; Brent Tarter, "'An Infant Borough entirely supported by Commerce': The Great Fire of 1776 and the Rebuilding of Norfolk," *VC* 28 (Autumn 1978): 52–61. See also Don Higginbotham, *The War of American Independence: Military Attitudes, Policies, and Practice, 1763–1789* (New York: Macmillan, 1971); John R. Alden, *A History of the American Revolution* (New York: Knopf, 1969); Samuel B. Griffith, *In Defense of the Public Liberty: Britain, America, and the Struggle for Independence—From 1760 to the Surrender at Yorktown in 1781* (Garden City, N.Y.: Doubleday, 1976); Merrill Jensen, *The Founding of a Nation: A History of the American Revolution, 1763–1776* (New York: Oxford University Press, 1968); George F. Scheer and Hugh F. Rankin, *Rebels and Redcoats* (New York: World, 1957); Claude H. Van Tyne, *The War of Independence, American Phase* (Boston: Houghton Mifflin, 1929); Hugh F. Rankin, *The American Revolution* (New York: Putnam, 1964); Hugh F. Rankin, *The War of the Revolution in Virginia* (Williamsburg: Virginia Independence Bicentennial Commission, 1979); John R. Alden, *The South in the Revolution, 1763–1789* (Baton Rouge: Louisiana State University Press, 1957); Charles M. Andrews, *The Colonial Background of the American Revolution,* rev. ed. (New Haven: Yale University Press, 1931); Mark Edward Lender, *A Respectable Army: The Military Origins of the Republic, 1763–1789* (Arlington Heights, Ill.: Harlan Davidson, 1982).

13.   Paul K. Walker, *Engineers of Independence: A Documentary History of the Army Engineers in the American Revolution, 1775–1783* (Washington, D.C.: OCE, 1981), 253–54; "Military Recommendations to the Governor and Council, May 15, 1778," *VMHB* 30 (1922): 187.

14.   Wertenbaker, *Norfolk,* 76–77; U.S. Writers' Program, *Virginia Guide,* 254–55; Emory G. Evans, *Thomas Nelson of Yorktown: Revolutionary Virginian* (Williamsburg, Virginia: Colonial Williamsburg Foundation, 1975); Emory G. Evans, "The Nelsons: A Biographical Study of a Virginia Family in the Eighteenth Cen-

tury" (Ph.D. diss., University of Virginia, 1957); David A. Clary, *The Inspectors General of the United States Army: A History, 1775–1903* (Bloomington, Ind.: David A. Clary and Associates, 1983), 186–87; David A. Clary and Joseph W. A. Whitehorne, *The Inspectors General of the United States Army, 1777–1903* (Washington, D.C.: DA, 1987), pt. 1.

15.   Mathew to General Sir Henry Clinton, 16 May, 24 May 1779, in "Expedition to Portsmouth, Virginia, 1779," *WMQ*, 2d ser., 12 (1972): 184–85; Wertenbaker, *Norfolk*, 76–77.

16.   Collier to Clinton, 16 May 1779, Mathew to Clinton, 24 May 1779, in "Expedition to Portsmouth, Virginia, 1779," 181–82, 185.

17.   Matthew Page Andrews, *Virginia: The Old Dominion* (Garden City, N.Y.: Doubleday Doran, 1937), 300–304; Arnold's report to Clinton, 12 May 1781, in "Arnold's Expedition to Richmond, Virginia, 1781," *WMQ*, 2d ser., 12 (1932): 187–90; Higginbotham, *War of American Independence*, 352–88; Francis Vinton Greene, *General Greene* (1893; rept. Port Washington, N.Y.: Kennikat, 1970), 168–76, 204–6; Theodore Thayer, *Nathanael Greene: Strategist of the American Revolution* (New York: Twayne, 1960), 23, 281–87, 290, 295–96; Rankin, *War of the Revolution in Virginia*, esp. 16–17, 21, 36–39, 41–43, 66, 73–74; Clary, *Inspectors General*, 183–93; Clary and Whitehorne, *Inspectors General*, 55–56; Vincent J. Esposito, *The West Point Atlas of American Wars*, 2 vols. (New York: Praeger, 1959); Lee A. Wallace, Jr., "The Battery at Hood's," *VC* 23 (Summer 1973): 38–47; Walker, *Engineers of Independence*, 263.

18.   Burke Davis, *The Campaign That Won America: The Story of Yorktown* (New York: Dial, 1970); Rankin, *War of Independence in Virginia;* Higginbotham, *War of American Independence*, 376–83.

19.   John W. Wright, "Notes on the Siege of Yorktown in 1781 with Special Reference to the Conduct of a Siege in the Eighteenth Century," *WMQ*, 2d ser., 12 (1932): 229–50; Paul K. Walker, "An Engineering Victory: The Siege of Yorktown, 1781," *ME* 73 (1981): 334–37; James N. Haskett, "Military Engineers at Yorktown, 1781," ibid., 68 (May–June 1976); Edward M. Riley, "Yorktown during the Revolution, Part II," *VMHB* 57 (1949): 176–88, 274–85; Charles E. Hatch, Jr., *Yorktown and the Siege of 1781* (Washington, D.C.: GPO, 1957).

20.   Heitman, 1:42–43; United States, *Journals of the Continental Congress, 1774–1789*, 34 vols. (Washington, D.C.: GPO, 1934–37), 16 June 1775, 27 May 1778, 11 March 1779; Raphael P. Thian, *Legislative History of the General Staff of the Army of the United States (Its Organization, Duties, Pay, and Allowances), from 1775 to 1901*, S. Doc. 229, 56 Cong., 2 sess. (1901), 485–95. See also Gilbert Bodinier, *Dictionnaire des Officiers de l'armée royale qui ont combattu aux Etats-Unis pendant la guerre d'Independance, 1776–1783* (Vincennes: Bureau d'histoire, 1982). The Corps of Engineers was not a formal bureaucratic organization in the Continental army, as was its successor in the nineteenth century. The sappers and miners formed a distinct unit commanded by the chief engineer of the main army, but mostly engineering was the responsibility of engineers (members of the Corps of Engineers) attached as staff officers to army or division commanders. Robert K. Wright, Jr., *The Continental Army* (Washington, D.C.: DA, 1983), 132; Louis C. Hatch, *The Administration of the American Revolutionary Army* (New York: Longmans Green, 1904); Rudolph Cronau, *The Army of the American Revolution and Its Organization* (New York: Cronau, 1923); E. Wayne Carp, "Supplying the Revolution: Continental Army Administration and American Political Culture, 1775–1783" (Ph.D. diss., University of California at Berkeley, 1981).

21. Arthur G. Peterson, "Commerce of Virginia, 1789–1791," *WMQ*, 2d ser., 10 (1930): 302–9. See also George D. Whitehurst, "The Commerce of Virginia, 1789–1815" (M.A. thesis, University of Virginia, 1951); Curtis P. Nettels, *The Emergence of a National Economy, 1775–1815* (New York: Holt, Rinehart and Winston, 1962); Douglass C. North, *The Economic Growth of the United States, 1790–1860* (New York: Norton, 1966).

22. Fillmore Norflett, trans. and ed., "Norfolk, Portsmouth, and Gosport as Seen by Moreau de Saint-Mery, in March, April, and May 1794," *VMHB* 48 (1940): 12–20, 153–64, 253–64; Wertenbaker, *Norfolk,* 97–98; Winston C. Babb, "French Refugees from Saint Domingue in the Southern United States, 1791–1810" (Ph.D. diss., University of Virginia, 1954).

23. Wertenbaker, *Norfolk,* 96–108; Tarter, "An Infant Borough Entirely Supported by Commerce," 58; William M. E. Rachal, "When Virginia Owned a Shipyard: The Story of the Norfolk Naval Shipyard at Portsmouth to the Time of Its Purchase by the United States in 1801," *VC* 2 (Autumn 1952): 31–35; Marshall Smelser, *The Congress Founds the Navy* (Notre Dame, Ind.: University of Notre Dame Press, 1959); Edward L. Beach, *The United States Navy: 200 Years* (New York: Holt, 1986), 1–50; Leonard D. White, *The Jeffersonians: A Study in Administrative History, 1801–1829* (New York: Macmillan, 1959).

ⅬⅬⅬⅬⅬⅬⅬⅬⅬⅬⅬ   2   ⅬⅬⅬⅬⅬⅬⅬⅬⅬⅬⅬ

1. Knox to the President, 18 Jan. 1790, quoted in James Ripley Jacobs, *The Beginning of the U.S. Army, 1783–1812* (Princeton, N.J.: Princeton University Press, 1947), 44; Weigley, 89–90; Richard H. Kohn, *Eagle and Sword: The Federalists and the Creation of the Military Establishment in America, 1783–1802* (New York: Free Press, 1975); Francis Paul Prucha, *The Sword of the Republic: The United States Army on the Frontier, 1783–1846* (Toronto: Macmillan, 1969); Harry M. Ward, *The Department of War, 1781–1795* (Pittsburgh: University of Pittsburgh, 1962). See also Tommy R. Young II, "The United States Army in the South, 1789–1835" (Ph.D. diss., Louisiana State University, 1973), Lawrence Delbert Cress, *Citizens in Arms: The Army and the Militia in American Society to the War of 1812* (Chapel Hill: University of North Carolina Press, 1982); Clary, *Inspectors General,* 214–57; Clary and Whitehorne, *Inspectors General,* 61–88; William H. Gaines, Jr., "The Forgotten Army: Recruiting for a National Emergency (1799–1800)," *VMHB* 56 (1948): 267–69; Carlos Emmor Godfrey, "Organization of the Provisional Army of the United States in the Anticipated War with France, 1798–1800," *PMHB* 38 (1914): 129–82; A. D. Schenck, "The United States Army in the Year 1801," *JMSIUS* 32 (1911): 443–44; Donald R. Hickey, "Federalist Defense Policy in the Age of Jefferson, 1801–1812," *MA* 45 (April 1981): 63–70; Smelser, *Congress Founds the Navy;* Marshall Smelser, "The Passage of the Naval Act of 1794," *MA* 22 (Winter 1958): 1–12.

2. Norflett, "Norfolk, Portsmouth, and Gosport as Seen by Moreau de Saint-Mery," 254–55.

3. Ibid., 28, 30, 255.

4. U.S. Congress, House Committee on Fortifications, *Fortifications,* MAD 13 (1794), *ASP-MA,* 1.

5. Lewis, 21–22; Edgar B. Wesley, "The Beginnings of Coast Fortifications,"

CAJ 67 (1927): 283–84. USWD, *Fortifications,* MAD 22 (19 Dec. 1794), *ASP-MA,* 1, lists the engineers and prints their reports.

6. *Fortifications,* MAD 22; Heitman, 1:833.

7. Knox to Governor of Virginia, 28 March 1794, Knox, Instructions to John Jacob Ulrich Rivardi, Acting as Temporary Engineer in the Service of the United States, 28 March 1794, *CVSP,* 7:87, 92–95. Identical copies of most of Rivardi's instructions and correspondence appear in *CVSP* and MAD 22; the former is cited hereafter, unless a document appears only in the latter.

8. Rivardi to Gov. of Va., 3 April, 15 April 1794, *CVSP,* 7:91–92, 110–11. See also Thomas Boyd, *Light-Horse Harry Lee* (New York: Harper, 1931).

9. Rivardi to the President, 6 May 1794, MAD 22. See also William P. Bradshaw, Jr., *Fort Norfolk Then and Now,* research by Julian Tompkins (Norfolk, n.d.), 6.

10. Knox to Lee, 9 May, 9 June 1794, *CVSP,* 7:138, 174.

11. SecTreasury to SecWar, 7 July 1794, SecWar to SecTreasury, 9 July 1794, MAD 22; Rivardi to Gov. of Va., 9 June 1794, *CVSP,* 7:173–74.

12. Rivardi to Gov. of Va., 15 June 1794, *CVSP,* 7:184–85.

13. Rivardi to SecWar, 24 June 1794, MAD 22.

14. Ibid., 6 July 1794.

15. Rivardi to Gov. of Va., 11 July 1794, *CVSP,* 7:212–13.

16. Ibid., 19 July 1794, pp. 221–22.

17. Rivardi to SecWar, 20 July 1794, MAD 22.

18. Knox to "Sir," 24 July 1794, ibid.

19. Knox to Gov. of Va., 30 July 1794, *CVSP,* 7: 237; Hamilton to Knox, 17 Dec. 1794, MAD 22; Assistant Chief of Engineers for Real Estate to Undersecretary of the Army, 21 June 1954, copy in 228–10 Installation Historical File, ND, Corps of Engineers.

20. House Committee on Fortifications, *Fortifications,* MAD 19 (4 Dec. 1794), and *Fortifications,* MAD 24 (28 Jan. 1795), both in *ASP-MA,* 1.

21. Rivardi to SecWar, 8 Dec. 1794, MAD 22.

22. See, for example, B. Green to Gov. of Va., 14 May 1795, *CVSP,* 7:502–3.

23. Pickering to President of the Senate, 16 Jan. 1796, in USWD, *State of the Fortifications of the United States,* MAD 26 (1796), *ASP-MA,* 1.

24. House Committee on Fortifications, *Fortifications,* MAD 28 (9 May 1796), *Fortifications,* MAD 29 (10 Feb. 1797), and *Fortifications,* MAD 31 (10 June 1797), all in *ASP-MA,* 1.

25. William A. Ganoe, *History of the United States Army,* rev. ed. (New York: Appleton-Century, 1942), 108; Weigley, 98–104; Jacobs, *Beginning of the U.S. Army,* 222–23; Godfrey, "Organization of the Provisional Army of the United States," 129–82; Gaines, "Forgotten Army," 267–69; Schenck, "United States Army in 1801," 443–44; Hickey, "Federalist Defense Policy," 63–70; Kohn, *Eagle and Sword,* 252–54; Clary, *Inspectors General,* 268–306; Clary and Whitehorne, *Inspectors General,* 73–83; Freeman, *George Washington,* 7:521–34; Stephen G. Kurtz, *The Presidency of John Adams: The Collapse of Federalism, 1795–1800* (Philadelphia: University of Pennsylvania Press, 1957), 308–33; Page Smith, *John Adams,* 2 vols. (Garden City, N.Y.: Doubleday, 1962), 2:973–1007, 1033–34; Bern-

hard Knollenberg, "John Adams, Knox, and Washington," *PAAS* 56 (Oct. 1946): pt. 2, pp. 20–38.

26. *Reorganization of the Army,* MAD 35 (1798), *ASP-MA,* 1; Stephen E. Ambrose, *Duty, Honor, Country: A History of West Point* (Baltimore: Johns Hopkins University Press, 1966): Sidney Forman, *West Point: A History of the United States Military Academy* (New York: Columbia University Press, 1950); Jacobs, *Beginning of the U.S. Army,* 298–99; Sidney Forman, "Why the United States Military Academy Was Established in 1802," *MA* 29 (Spring 1965): 16–28; Theodore J. Crackel, "Jefferson, Politics, and the Army: An Examination of the Military Peace Establishment Act of 1802," *JER* 2 (Spring 1982): 21–38; *Military Academy,* MAD 40 (1800), and *Military Academy, and Reorganization of the Army,* MAD 39 (1800), both in *ASP-MA,* 1.

27. See Sidney Forman, "The United States Military Philosophical Society, 1820–1813," *WMQ,* 3d ser., 2 (1945): 273–85; Arthur P. Wade, "A Military Offspring of the American Philosophical Society," *MA* 38 (Sept. 1974): 103–7.

28. McHenry to Samuel Sewall, 27 Feb. 1798, and the committee report and other documents in House Committee on Fortifications, *Fortifications,* MAD 32 (8 March 1798), and *Fortifications, Munitions, and Increase of the Army,* MAD 33 (1798), both in *ASP-MA,* 1.

29. Willard B. Robinson, *American Forts: Architectural Form and Function* (Urbana: University of Illinois Press, 1977), 67.

30. McHenry, 5 Jan. 1800, quoted in *Military Academy, and Reorganization of the Army,* MAD 39 (1800), *ASP-MA,* 1.

31. McHenry to Chairman, Committee on Defense, 1 May 1800, in House Committee on Defense, *Fortifications,* MAD 43 (1800), *ASP-MA,* 1; Wertenbaker, *Norfolk,* 112–13; *Military Force, the Posts at Which Stationed, and the Expenses of Fortifications, Arsenals, Armories, and Magazines, in the Years 1803 and 1804,* MAD 55 (1805); *Reduction of the Army Considered,* MAD 168 (1818), and *Fortifications,* MAD 60 (1806), all in *ASP-MA,* 1.

32. Dearborn to Roger Nelson, 9 Dec. 1806, in House Committee on Defense, *Fortifications and Gunboats,* MAD 66 (1807), and Senate, *Fortifications and Gunboats,* MAD 64 (1806), both in *ASP-MA,* 1.

33. Beach, *United States Navy,* 51–71; Edwin M. Gaines, "The *Chesapeake* Affair: Virginians Mobilize to Defend National Honor," *VMHB* 64 (1956): 131–42; Edwin M. Gaines, "Outrageous Encounter: The *Chesapeake-Leopard* Affair of 1807" (Ph.D. diss., University of Virginia, 1960); Wertenbaker, *Norfolk,* 110–11; Bradshaw, *Fort Norfolk,* 14–19. See also John C. Emmerson, *The Chesapeake Affair of 1807* (Portsmouth, Va.: Printercraft Press, 1954); Thomas P. Abernethy, *The South in the New Nation, 1789–1819* (Baton Rouge: Louisiana State University Press, 1961), 314–19.

34. Lewis, 25.

35. USWD, *Fortifications,* MAD 76 (8 Dec. 1807), Dearborn to Samuel L. Mitchill, 20 Nov. 1807, in Senate, *Fortifications and Gunboats,* MAD 74 (3 Dec. 1807), and House Committee on Defense, *Fortifications and Gunboats,* MAD 72 (24 Nov. 1807), all in *ASP-MA,* 1.

36. Knox to Mitchill, 20 Nov. 1807, MAD 74.

37. Thomas Jefferson to Senate and House of Representatives, 6 Jan. 1809, in *Fortifications,* MAD 84 (6 Jan. 1809), *ASP-MA,* 1; *Lands and Buildings Acquired for Military Purposes,* MAD 203 (1821), *ASP-MA,* 2; Bradshaw, *Fort Norfolk,* 20–21.

38. Jefferson to Senate and House of Representatives, 6 Jan. 1809, MAD 84; Wallace, "The Battery at Hood's," 38–47.

39. USWD, *Fortifications,* MAD 89 (1809), *ASP-MA,* 1.

40. House Committee on Defense, *Fortifications,* MAD 106 (1811), *ASP-MA,* 1.

41. Lewis, 26.

42. Forman, "United States Military Philosophical Society," 284; Heitman, 1:169.

43. Armstrong to Senator Joseph Anderson, 10 June 1813, in *Additional Fortifications, and an Increase of the Army,* MAD 124 (1813), *ASP-MA,* 1; Glenn Tucker, *Poltroons and Patriots: A Popular Account of the War of 1812,* 2 vols. (Indianapolis: Bobbs-Merrill, 1954), 1:299; Weinert and Arthur, 21–22; J. Mackay Hitsman and Alice Sorby, "Independent Foreigners or Canadian Chasseurs," *MA* 25 (Spring 1961): 11–17; Parke Rouse, Jr., "Low Tide at Hampton Roads," *USNIP* 95 (July 1969): 77–86; Robert Leckie, *The Wars of America,* 2 vols. (New York: Harper and Row, 1968), 1:254–57; Wertenbaker, Norfolk, 121–24; Starkey, *First Plantation,* 54–58; "Norfoliensis," "The Defence of Craney Island," *VHR* 1 (1848): 132–41; William H. Gaines, Jr., "Craney Island, or Norfolk Delivered," *VC* 1 (Winter 1951): 32–36. See also Henry Adams, *History of the United States during the Administrations of Jefferson and Madison,* 9 vols. (New York: Scribner, 1889–91); Henry Adams, *The War of 1812,* ed. Henry A. DeWeerd (Washington, D.C.: Infantry Journal, 1944); Leckie, *Wars of America,* 1:217–314; J. C. A. Stagg, *Mr. Madison's War: Politics, Diplomacy, and Warfare in the Early American Republic, 1783–1830* (Princeton, N.J.: Princeton University Press, 1983); John K. Mahon, *The War of 1812* (Gainesville: University of Florida Press, 1972); Reginald Horsman, *The War of 1812* (Chicago: University of Chicago Press, 1965); Gilbert Byron, *The War of 1812 on the Chesapeake Bay* (Baltimore: Maryland Historical Society, 1964); Edgar Erskine Hume, "Letters Written during the War of 1812 by the British Naval Commander in American Waters (Admiral Sir David Milne)," *WMQ,* 2d ser., 10 (1930): 279–301.

44. "The Vigilance Committee: Richmond during the War of 1812," *VMHB* 7 (Jan. 1900): 227–28, 232–33; Hitsman and Sorby, "Independent Foreigners," 11–17; Rouse, "Low Tide," 77–86; Weinert and Arthur, 22; Leckie, *Wars of America,* 1:256; Tucker, *Poltroons,* 1:299; Gillie Cary McCabe, *The Story of an Old Town: Hampton, Virginia* (Richmond: Old Dominion Press, 1929).

45. See "Richmond during the War of 1812: Letters of Dr. Thomas Massie," *VMHB* 7 (April 1900): 406–18.

46. "Vigilance Committee," 235–37, 414, 427; Wallace, "Battery at Hood's," 38–47. See also Adams, *War of 1812,* 218–30; Coles, *War of 1812,* 182–85, 248–50; Leckie, *Wars of America,* 1:279–91; Walter Lord, *The Dawn's Early Light* (New York: Norton, 1972).

⎍⎍⎍⎍⎍⎍⎍⎍⎍⎍ **3** ⎍⎍⎍⎍⎍⎍⎍⎍⎍⎍

1. H. Lallemand, *A Treatise on Artillery,* 2 vols. (New York: C. W. Van Winkle, 1820), 2:103–4.

2. Lewis, 37–38; Jamie W. Moore, *The Fortifications Board, 1816–1826, and the Definition of National Security* (Charleston, S.C.: The Citadel Press, 1981); Jamie W. Moore, "National Security in the American Army's Definition of Mission, 1865–

1914," *MA* 46 (Oct. 1982): 128. See also William H. Carter, "Bvt. Maj. Gen. Simon Bernard," *JMSIUS* 51 (1912): 147–55; Heitman, 1:214; Christopher Duffy, *Fire and Stone: The Science of Fortress Warfare, 1660–1860* (New York: Hippocrene Books, 1975); Isaac Malthy, *The Elements of War* (Boston: Thomas B. Waite, 1811); William B. Skelton, "The Commanding General and the Problem of Command in the United States Army 1821–1841, *MA* 34 (Dec. 1970): 117–22; Carlton B. Smith, "Congressional Attitudes toward Military Preparedness during the Monroe Administration," ibid., 40 (Feb. 1976): 22–25; David Ted Childress, "The Army in Transition: The United States Army, 1815–1841 (Ph.D. diss., Mississippi State University, 1974); William Barrett Skelton, "The United States Army, 1821–1837: An Institutional History" (Ph.D. diss., Northwestern University, 1968); Carlton B. Smith, "The United States War Department, 1815–1842" (Ph.D. diss., University of Virginia, 1967).

3. Heitman, 1:682, 966; *Dictionary of American Biography,* s.v. "Elliott, Jesse Duncan"; Beach *United States Navy,* 121–23, 143–41, 192–93.

4. These first three American phases of fort building should not be confused with the first, second, and third "systems" developed by the French godfather of fort design, Vauban.

5. USWD, *Fortifications,* MAD 206 (15 Feb. 1821, including the first full report of the Board of Engineers, 7 Feb. 1821), *ASP-MA,* 2; Swift to SecWar, 12 Oct. 1818, in USWD, *Estimates for the Year 1819,* MAD 169 (1818), *ASP-MA,* 1. Original records of the Board of Engineers for Fortifications may be found in Correspondence Relating to Fortifications and Records of Engineer Boards and Commissions (Board of Engineers for Fortifications, 1825–30), in ROCE.

6. *Reduction of the Army,* MAD 197 (1820), *ASP-MA,* 2; Heitman, 2:580–81.

7. USWD, *Fortifications,* MAD 206.

8. *Condition of the Military Establishment and the Fortifications, and Returns of the Militia,* MAD 247 (1823), and *Armament of Fortifications,* MAD 242 (1823), both in *ASP-MA,* 2; ARSecWar 1825, and ARCE 1825, both in MAD 284, *ASP-MA,* 3; *Revised Report of the Board of Engineers on the Defence of the Seaboard,* included in *On the Subject of an Augmentation of the Corps of Engineers of the Army,* MAD 327 (1826), *ASP-MA,* 3; ARSecWar 1828, MAD 390, *ASP-MA,* 4; *On the Necessity of an Increase of the Engineer Corps and Topographical Engineers Exclusively for Military Purposes,* MAD 463 (1831), *ASP-MA,* 4; *On the Importance of the Topographical Engineers of the Army,* MAD 465 (1831), *ASP-MA,* 4; ARSecWar 1834, MAD 585, and ARCE 1835, MAD 613, both in *ASP-MA,* 5. See also Richard L. Watson, Jr., "Congressional Attitudes toward Military Preparedness, 1829–1835," *MVHR* 34 (March 1948): 611–36.

9. *On the Means Necessary for the Military and Naval Defenses of the Country,* MAD 671, 24 Cong., 1 sess. (1836), *ASP-MA,* 6.

10. Cass did try to get more engineers. Cass to Thomas Hart Benton, 14 Jan. 1836, in *Recommendation for an Increase of the Corps of Engineers,* MAD 623 (1836), *ASP-MA,* 6. The Senate asked in April just how much the Engineer and Ordnance Departments could spend in a year. The engineers said $6 million, ordnance $3 million; they got neither. *Statement of the Maximum Amount That Can Be Expended Annually upon the Construction of Fortifications and the Ordnance Department,* MAD 674 (1836), *ASP-MA,* 6. See also *On the Number of Cannon Required to Arm the Fortifications, the Number That Can Be Supplied in a Year, and on the Expediency of Establishing a National Foundery,* MAD 680 (1836), *ASP-MA,* 6.

11.   Joseph Gardner Swift, *The Memoirs of Gen. Joseph Gardner Swift, LL.D., U.S.A., First Graduate of the United States Military Academy, West Point, Chief Engineer U.S.A. from 1812 to 1818* (n.p.: privately printed, 1890), 180.

12.   Quoted in Weinert and Arthur, 35. Fort Monroe covered the largest acreage of any Third System fort, but Fort Jefferson, Florida, was built (1850s) for the most guns (420) and involved a greater mass of masonry.

13.   ARSecWar 1836 and ARCE 1836, both in MAD 699, *ASP-MA*, 6; ARSecWar 1838, S. Doc. 1, 25 Cong., 3 sess., 98, 103.

14.   ARCE 1839, S. Doc. 1, 26 Cong., 1 sess., 157; ARCE 1840, S. Doc. 1, 26 Cong., 2 sess., 99; ARSecWar 1841, S. Doc. 1, 27 Cong., 2 sess., 62–64; Moore, "National Security," 128; ARSecWar 1842, S. Doc. 1, 27 Cong., 3 sess., 179, 183.

15.   ARComGen 1843, ARSurGen 1843, and ARQMG 1843, all in S. Doc. 1, 28 Cong., 1 sess., 64, 72–73, 81–82; Weigley, 168. See also David A. Clary, *These Relics of Barbarism: A History of Furniture in Barracks, Hospitals, and Guardhouses of the United States Army, 1800–1880* (Bloomington, Ind.: David A. Clary and Associates, 1982; rept. in part Harpers Ferry, W.Va.: National Park Service, 1985); David A. Clary, *A Life Which Is Gregarious in the Extreme: A History of Furniture in Barracks, Hospitals, and Guardhouses of the United States Army, 1880–1945* (Bloomington, Ind.: David A. Clary and Associates, 1983); Erna Risch, *Quartermaster Support of the Army: A History of the Corps, 1775–1939* (Washington, D.C.: DA, 1962), 210.

16.   ARSecWar 1844, 115, and ARCE 1844, 188–90, both in S. Doc. 1, 28 Cong., 2 sess.; ARCE 1845, S. Doc. 1, 29 Cong., 1 sess., 266–69; ARCE 1846, S. Doc. 1, 29 Cong., 2 sess., 133; ARCE 1847, S. Ex. Doc. 1, 30 Cong., 1 sess., 623; ARCE 1851, S. Ex. Doc. 1, 32 Cong., 1 sess., 357–58.

17.   ARSecWar 1844, 118–19; ARSecWar 1845, S. Doc. 1, 29 Cong., 1 sess., 200–201; ARSecWar 1846, S. Doc. 1, 29 Cong., 2 sess., 56–57.

18.   ARSecWar 1847, S. Ex. Doc. 1, 30 Cong., 1 sess., 67; ARCE 1847, 594–96; ARSecWar 1848, 81–82, and ARCE 1848, 247, both in H. Ex. Doc. 1, 30 Cong., 2 Sess.; ARCE 1849, S. Ex. Doc. 1, 31 Cong., 1 sess., 209–11. See also Francis Paul Prucha, "Distribution of Regular Army Troops before the Civil War," *MA* 16 (Winter 1952): 169–73.

19.   Moore, "National Security," 128; ARSecWar 1851, 105, 109, and ARCE 1851, 344–45, 361, both in S. Ex. Doc. 1, 31 Cong., 2 sess., pt. 2; ARSecWar 1852, 6–8, and ARCE 1852, 147, both in S. Ex. Doc. 1, 32 Cong., 2 sess.; ARSecWar 1853, S. Ex. Doc. 1, 33 Cong., 1 sess., pt. 2:5.

20.   ARSecWar 1853, 15–16; ARCE 1853, S. Ex. Doc. 1, 33 Cong., 1 sess., pt. 2:157–63.

21.   ARCE 1853, 201–3; ARSecWar 1854, 5, and ARCE 1854, 92–98, both in S. Ex. Doc. 1, 33 Cong., 3 sess., pt. 2; ARCE 1855, 190–91; ARSecWar 1856, 15–16, and ARCE 1856, 272–73, both in S. Ex. Doc. 5, 34 Cong., 3 sess., pt. 2; ARSecWar 1857, 16–17, and ARCE 1857, 168, both in S. Ex. Doc. 11, 35 Cong., 1 sess., pt. 2; Moore, "National Security," 128.

22.   Moore, "National Security," 128; ARSecWar 1859, S. Ex. Doc. 2, 36 Cong., 1 sess., pt. 2:11–12.

23.   ARSecWar 1859, 452–579. For Morton, see Heitman, 1:731.

24.   ARCE 1859, S. Ex. Doc. 2, 36 Cong., 1 sess., pt. 2:634–35; ARCE 1860, S. Ex. Doc. 1, 36 Cong., 2 sess., pt. 2:253; Moore, "National Security," 128.

25. Weinert and Arthur, 27.

26. USWD, *Contracts Made in the Year 1818,* MAD 175 (1819), *ASP-MA,* 1; Robert Meriwether, W. Edwin Hemphill, et al., eds., *The Papers of John C. Calhoun* (Columbia: University of South Carolina Press, 1959—), 2:1xi; Macomb to SecWar, 30 April 1822, in House Special Committee on Contract with Elijah Mix, *Contract for Stone at the Rip Raps and Old Point Comfort,* MAD 234 (1822), *ASP-MA,* 2.

27. Swift, *Memoirs,* 189; *Contract for Stone at the Rip Raps and Old Point Comfort;* William M. Meigs, *The Life of John Caldwell Calhoun,* 2 vols. (New York: Neale, 1917), 2:266–72; Gerald M. Capers, *John C. Calhoun—Opportunist: A Reappraisal* (Gainesville: University of Florida Press, 1960), 86n, 113; Margaret L. Coit, *John C. Calhoun, American Patriot* (Boston: Houghton Mifflin, 1950), 165.

28. Swift to Lewis, 3 April 1817, Swift to Colonel McRee, 1 Sept. 1817, General Simon Bernard to Colonel Armistead, 10 Sept. 1817, Swift to SecWar, 11 Sept. 1817 (2 letters), Buell Collection of Historical Documents Relating to the Corps of Engineers, 1801–19, ROCE, microcopy 417; Swift, *Memoirs,* 170; Calhoun to Swift, 6 Jan. 1818, Meriwether and Hemphill, *Papers of Calhoun,* 2:61.

29. *Fortifications,* MAD 218 (1822), *ASP-MA,* 2; Swift to SecWar, 10 and 29 Jan. 1818, Report of the Board of Engineers on Defenses of Hampton Roads, 24 Jan. 1818, and orders to Colonels Armistead and McRee and Majors I. Roberdeau and James Kearney, 28 Jan. 1818, Buell Collection, ROCE; Swift, *Memoirs,* 170–71; Swift to SecWar, 29 Jan. 1818, Meriwether and Hemphill, *Papers of Calhoun,* 2:61.

30. Swift to McRee, 1 and 15 May 1818, Buell Collection, ROCE; Swift *Memoirs,* 173–76; Weinert and Arthur, 27.

31. Swift to Calhoun, 12 Oct. 1818, in USWD, *Estimates for the Year 1819,* MAD 169 (1818), *ASP-MA,* 1.

32. Weinert and Arthur, 27; Armistead to Calhoun, 12 Jan. 1820, in *Fortifications,* MAD 183 (1820), *ASP-MA,* 2.

33. *Fortifications,* MAD 218 (1822), *ASP-MA,* 2; Weinert and Arthur, 30.

34. USWD, *Fortifications,* MAD 106 (1821), *ASP-MA,* 2.

35. *Lands and Buildings Acquired for Military Purposes,* MAD 203 (1821), *ASP-MA,* 2; *Fortifications,* MAD 183 (1820), *ASP-MA,* 2; *Reduction of the Army Considered,* MAD 168 (1818), *ASP-MA,* 1; *System of Fortifications Recommended by the Board of Engineers,* MAD 316 (1826), *ASP-MA,* 3; William McRee to W. T. Poussin, 15 Oct. 1818, Buell Collection, ROCE.

36. *Revised Report of the Board of Engineers on the Defence of the Seaboard,* in *On the Subject of an Augmentation of the Corps of Engineers of the Army,* MAD 327 (1826), *ASP-MA,* 3; Reports of the Adjutant General in *Condition of the Military Establishment and Fortifications,* MAD 235 (1822), and *Condition of the Military Establishment and the Fortifications, and Returns of the Militia,* MAD 247 (1823), both in *ASP-MA,* 2; *General Statement of Desertions and Death for Three Years Ending September 20, 1825,* MAD 326 (1826), *ASP-MA,* 3; Report of the Adjutant General in *Condition of the Military Establishment and the Fortifications,* MAD 262 (1824), *ASP-MA,* 2; *System of Fortifications Recommended by the Board of Engineers,* MAD 316 (1826), *ASP-MA,* 3; U.S. Department of the Interior, *The National Register of Historic Places, 1976* (Washington, D.C.: GPO, 1976), 805.

37. *On the Means Necessary for the Military and Naval Defenses of the Country,* MAD 671 (1836), *ASP-MA,* 6; James Smith to SecNavy, 20 July 1848, Totten to SecWar, 28 July 1848, and Robert M. McGuirt, Management and Disposal Branch, ND, to Joe Fehrer, Washington District, 14 Oct. 1954, copies of all in 228–10 Installation Historical File, ND.

38. *Condition of the Military Establishment* (1822); *Condition of the Military Establishment* (1823); *System of Fortifications Recommended by the Board of Engineers* (1826); *Condition of the Military Establishment* (1824); ARCE 1825; *Revised Report of the Board of Engineers on the Defence of the Seaboard* (1826); ARCE 1826, MAD 334, and ARCE 1827, MAD 360, both in *ASP-MA*, 3; ARCE 1828, MAD 390, ARCE 1829, MAD 410, ARCE 1820, MAD 458, and ARCE 1831, MAD 485, all in *ASP-MA*, 4; ARCE 1832, MAD 532, ARCE 1833, MAD 551, and ARCE 1834, MAD 585, all in *ASP-MA*, 5; *On the Means Necessary for the Military and Naval Defenses of the Country* (1836).

39. Macomb in *Condition of the Military Establishment* (1823); *On the Subject of an Augmentation of the Corps of Engineers of the Army,* MAD 327 (1826), *ASP-MA*, 3; ARCE 1829.

40. ARCE 1829; ARCE 1830. For Talcott, Dutton, Mansfield (John Mansfield of Maryland or Joseph K. F. Mansfield), and Gratiot, see Heitman, 1:391, 470, 689, 943.

41. Weinert and Arthur, 69–70; ARCE 1835. For Lee and Eliason, see Heitman, 1:401, 625. See also George G. Shackleford, ed. "Lieutenant Lee Reports to Captain Andrew Talcott on Fort Calhoun's Construction on the Rip Raps," *VMHB* 60 (1952): 458–87.

42. *Recommendation of Appropriations for the Construction and Armament of Fortifications for the National Defense,* MAD 626 (1836), *ASP-MA*, 6; ARCE 1836; *The Number and Cost of Fortifications, Arsenals, and Armories Completed and in Progress,* MAD 701 (1836), *ASP-MA*, 6; *Statements of the Fortifications, Their Garrisons, Steam Batteries, Etc., Necessary for the Defense of the Coasts of the United States, Their Costs, Etc.,* MAD 650 (1836), *ASP-MA*, 6; ARCE 1837, MAD 745, *ASP-MA*, 7; ARCE 1838, S. Doc. 1, 25 Cong., 3 sess., 338–41; ARCE 1839, 236–37, 240–41; ARCE 1840, 160–61; ARCE 1854, 111–12; ARCE 1855, S. Ex. Doc. 1, 34 Cong., 1 sess., pt. 2:209; ARCE 1856, 280–81; ARCE 1857, 176. For Smith, see Heitman, 1:900.

43. ARCE 1858, S. Ex. Doc. 1, 35 Cong., 2 sess., pt. 3:820–21; ARCE 1860, 271. For DeRussy, see Heitman, 1:369.

44. Robinson, *American Forts,* 99; Weinert and Arthur, 28–29.

45. Weinert and Arthur, 31; Register of Work Done by Slave Labor at Fort Monroe, 1821–24, Time Roll for Fort Monroe, 1826–28, and RO at Forts Monroe and Calhoun, 1830–32, all in Records of Fort Monroe, 1815–1917, Records of the Norfolk District, ROCE.

46. Armistead to SecWar, 10 Feb. 1821, in USWD, *Fortifications,* MAD 206 (1821), *ASP-MA*, 2; *Condition of the Military Establishment* (1823).

47. *Fortifications,* MAD 155 (1824), *ASP-MA*, 2.

48. *Condition of the Military Establishment* (1824); Wertenbaker, *Norfolk,* 135.

49. Registers of Stone Received and Registers of Materials Received at Fort Monroe and Fort Calhoun, 1826–27, Records of Fort Monroe, ROCE; ARCE 1825; ARCE 1826; ARCE 1827.

50. Cecil D. Eby, Jr., ed., "Recollections of Fort Monroe, 1826–1828: From Autobiography of Lieutenant Alfred Beckley," *VMHB* 72 (1964): 479–89.

51. Registers of Stone Received at Forts Monroe and Calhoun, 1827–28, RO at Forts Monroe and Calhoun, 1830–32, and Registers of Material Received, 1829–31, Records of Fort Monroe, ROCE; ARCE 1828; ARCE 1829; ARCE 1831.

52. "Fort George," *VHR* 1, no. 1 (1848): 22; RO at Forts Monroe and Calhoun, 1830–32, and Daily RO at Forts Monroe and Calhoun, 1832, Records of Fort Monroe, ROCE; ARCE 1832.

53. Weinert and Arthur, 63; Daily RO at Forts Monroe and Calhoun, 1833–34, Records of Fort Monroe, ROCE; ARCE 1833; ARCE 1834.

54. ARCE 1836; ARSecWar 1837, MAD 745, *ASP-MA*, 7; ARCE 1837.

55. Daily RO at Fort Monroe, 1838, and Register of Expenditures for Bridges, Contingencies, and Repairs of Roads, 1837–38, Records of Fort Monroe, ROCE; ARCE 1837; ARCE 1838, 181; ARCE 1839, 165; ARSecWar 1838, S. Ex. Doc. 1, 25 Cong., 3 sess., 104.

56. Daily RO at Fort Monroe, 1839, Register of Expenditures for Bridges, Contingencies, and Repairs of Roads, 1838–39, and Eliason to Chief of Engineers, 1 July 1839, Letters Sent, Records of Fort Monroe, ROCE; ARCE 1839, 165–66.

57. ARCE 1840, 104, and ARSecWar 1840, 20, both in S. Doc. 1, 26 Cong., 2 sess.

58. Daily RO at Fort Monroe, 1841–42, Records of Fort Monroe, ROCE; ARCE 1841, S. Doc. 1, 27 Cong., 2 sess., 122–23; ARCE 1842, S. Doc. 1, 27 Cong., 3 sess., 253.

59. Daily RO at Fort Monroe, 1841–47, and Register of Expenditures for Bridges, Contingencies, and Repairs of Roads, 1841–47, Records of Fort Monroe, ROCE; ARCE 1842, 253; ARCE 1843, S. Doc. 1, 28 Cong., 1 sess., 103–4; ARCE 1844, 175–76; ARCE 1845, 252–53; ARCE 1846, 122–23; ARCE 1847, 609–10; ARCE 260–61; RO, 1848–51, and Record of Expenditures for Bridges, Contingencies, and Repairs of Roads, 1848, Records of Fort Monroe, ROCE; ARCE 1849, 217–18; ARCE 1850, S. Ex. Doc. 1, 31 Cong., 2 sess., pt. 2:355; ARCE 1851, 251–52; ARCE 1852, 153; ARCE 1853, 168; ARCE 1854, 103.

60. RO, 1855–56, and Smith to Chief of Engineers, 30 June 1856, Records of Fort Monroe, ROCE; ARCE 1855, 197–98; ARCE 1856, 280–81.

61. RO, 1857–61, Records of Fort Monroe, ROCE; ARCE 1857, 177; ARCE 1858, 819–20; ARCE 1859, 644; ARCE 1860, 262; ARCE 1861, S. Ex. Doc. 1, 37 Cong., 2 sess., pt. 2:100. See also Richard P. Weinert, Jr., *The Guns of Fort Monroe* (Fort Monroe, Va.: Fort Monroe Casemate Museum, 1974).

62. Richard P. Weinert, Jr., "Saga of Old Fort Wool," *PJCAMP* 8 (Winter 1976–77): 3–14.

63. Armistead to Calhoun, 10 Feb. 1821, in USWD, *Fortifications,* MAD 206 (1821), *ASP-MA,* 2. Fort Calhoun's incomplete construction records are filed along with those of Fort Monroe in ROCE.

64. *Condition of the Military Establishment* (1823); *Fortifications,* MAD 255 (1824), *ASP-MA,* 2; *Condition of the Military Establishment* (1824).

65. ARCE 1825; Registers of Stone Received at Forts Monroe and Calhoun, 1817–28, Records of Fort Monroe, ROCE.

66. William E. Beard, "The Castle of Rip Raps," *CAJ* 78 (1935): 44–48; Weinert, "Saga of Old Fort Wool," 3.

67. ARCE 1826; ARCE 1827; ARCE 1828; Record of Material Received at Rip Rap Shoals, 1819–20, and Registers of Stone Received at Forts Monroe and Calhoun intermittently from 1817 to 1841, Records of Fort Monroe, ROCE.

68. RO at Forts Monroe and Calhoun, 1830–32, Records of Fort Monroe, ROCE; ARCE 1829; ARCE 1831.

69. ARCE 1832; ARCE 1833; ARCE 1834; RO at Forts Monroe and Calhoun, 1831–32, Records of Fort Monroe, ROCE.

70. Daily RO at Forts Monroe and Calhoun, 1834–35, Records of Fort Monroe, ROCE; ARSecWar 1835; ARCE 1835.

71. Daily RO at Fort Calhoun, 1835–39, Records of Fort Monroe, ROCE; ARCE 1836; *Explanations of the Estimates for Fortifications for the Year 1837,* MAD 706 (1836), *ASP-MA,* 6; ARSecWar 1837; ARCE 1837; ARCE 1838, 104, 181–82; ARCE 1839, 166.

72. Daily RO at Fort Calhoun, 1838–40, and Registers of Stone Received at Forts Monroe and Calhoun, 1839–40, Records of Fort Monroe, ROCE; ARCE 1839, 166; ARCE 1840, 104.

73. Daily RO at Fort Calhoun, 1840–42, and Registers of Stone Received at Forts Monroe and Calhoun, 1841, Records of Fort Monroe, ROCE; ARSecWar 1841, 64; ARCE 1841, 123; ARCE 1842, 253–54.

74. ARCE 1843, 104; ARSecWar 1844, 119–20, ARCE 1844, 176; ARCE 1845, 253–54.

75. ARCE 1846, 123; ARCE 1847, 610; ARCE 1848, 261–62; ARCE 1849, 218; ARCE 1850, 355–56; ARCE 1851, 352; ARSecWar 1852, 13–14; ARCE 1852, 153; ARCE 1853, 168; ARCE 1854, 104; ARCE 1855, 198; ARCE 1856, 281; ARCE 1857, 177; ARCE 1858, 820–21; ARCE 1859, 645; ARCE 1860, 262–63; ARCE 1861, 100; Daily RO at Fort Calhoun, 1855–61, Records of Fort Monroe, ROCE.

76. Lewis, 43–44; Quentin Hughes, *Military Architecture* (New York: St. Martin's Press, 1974), 178; Joseph G. Totten, *Report on the Effects of Firing with Heavy Ordnance from Casemate Embrasures [and] against the Same Embrasures,* Papers on Practical Engineering, no. 6 (Washington, D.C.: Taylor and Maury, 1857).

77. Colonel R. Delafield, *Report on the Art of War in Europe in 1854, 1855, and 1856,* H. Ex. Doc., 36 Cong., 2 sess. (Washington, D.C.: GPO, 1861).

78. ARSecWar 1861, S. Ex. Doc. 1, 37 Cong., 2 sess., pt. 2:8. Robert S. Browning III, in *Two If by Sea: The Development of American Coastal Defense Policy* (Westport, Conn.: Greenwood, 1983), asserts that the motivation for American coastal defenses from 1794 into the twentieth century was not to resist attack but to deter it. Cameron's statement, however, is the only important utterance in the nineteenth century that might support such a claim. Compare Secretary of War Lincoln's 1883 statement (see chap. 6 below) for a more typical line of thought.

79. ARCE 1861, 94–95; ARSecWar 1862, S. Ex. Doc. 1, 37 Cong., 3 sess., pt. 4:13–15; ARSecWar 1863, H. Ex. Doc. 1, 38 Cong., 1 sess., pt. 5:11; ARCE 1864, H. Ex. Doc. 83, 38 Cong., 2 sess., 30.

80. Lewis, 66–71.

81. ARSecWar 1864, H. Ex. Doc. 83, 38 Cong., 2 sess., 6–7; ARCE 1864, 32.

82. ARSecWar 1865, 42, and ARCE 1865, 918–19, both in H. Ex. Doc. 1, 39 Cong., 1 sess., pt. 3. For Delafield, see Heitman, 1:365.

🮔🮔🮔🮔🮔🮔🮔🮔🮔🮔 **4** 🮔🮔🮔🮔🮔🮔🮔🮔🮔🮔

1. Weinert and Arthur, 84. For Wright, see Heitman, 1:1062.

2. Ibid., 88–90, 102, 110.

3. Beach, *United States Navy,* 241–301; Bruce Catton, *Terrible Swift Sword* (Garden City, N.Y.: Doubleday, 1963), 204–15. See also James Phinney Baxter, *The Introduction of the Ironclad Warship* (Cambridge: Harvard University Press, 1933).

4. Quoted in Weinert and Arthur, 113–14. For Wool, see Heitman, 1:1059–60; Harwood P. Hinton, "The Military Career of John Ellis Wool, 1812–1863" (Ph.D. diss., University of Wisconsin, 1960); Clary, *Inspectors General,* 401–3, 428–597 passim.

5. Weinert and Arthur, 116–19.

6. Ibid., 122–23.

7. RO at Fort Wool 1864, Records of Fort Monroe, ROCE; Weinert and Arthur, 124.

8. Weigley, 150–51. See also Dennis Hart Mahan, *An Elementary Treatise on Advanced Guard, Out-Post, and Detached Service of Troops* (1847), rev. ed. (New York: Wiley, 1864); R. Ernest Dupuy, *Where They Have Trod: The West Point Tradition in American Life* (New York: Stokes, 1940), 228–40; R. Ernest Dupuy, *The Men of West Point: The First 150 Years of the United States Military Academy* (New York: Sloane, 1951), 12–24; Russell F. Weigley, *Towards an American Army: Military Thought from Washington to Marshall* (New York: Columbia University Press, 1962), 42–53.

9. Edward Hagerman, "From Jomini to Dennis Hart Mahan: The Evolution of Trench Warfare and the American Civil War," *CWH* 13 (Sept. 1967): 197–220. See also Jack Coggins, *Arms and Equipment of the Civil War* (Garden City, N.Y.: Doubleday, 1962), Frederick P. Todd, *American Military Equipage, 1851–1872* (Providence: Company of Military Historians, 1974).

10. John A. Carpenter, "O. O. Howard: General at Chancellorsville," *CWH* 3 (March 1957): 53–54.

11. Hagerman, "From Jomini to Dennis Hart Mahan," 219.

12. William T. Sherman, *Memoirs of William T. Sherman,* 2 vols. (1875; rept. Bloomington: Indiana University Press, 1957), 2:396–97.

13. *Letter of the Secretary of War, Transmitting Report on the Organization of the Army of the Potomac, and of Its Campaigns in Virginia and Maryland, under the Command of Maj. Gen. George B. McClellan, from July 26, 1861, to November 7, 1862,* H. Ex. Doc. 15, 38 Cong., 1 sess. (1864); Bruce Catton, *Mr. Lincoln's Army* (Garden City, N.Y.: Doubleday, 1951); Alexander S. Webb, *The Peninsula: Mc-Clellan's Campaign of 1862* (New York: Scribner, 1881); Stephen W. Sears, *Landscape Turned Red: The Battle of Antietam* (New York: Ticknor and Fields, 1983); Warren W. Hassler, Jr., *General George McClellan: Shield of the Union* (Baton Rouge: Louisiana State University Press, 1957); Joseph G. Barnard and W. F. Barry, *Report of the Engineer and Artillery Operations of the Army of the Potomac, from Its Organization to the Close of the Peninsular Campaign* (New York: Van Nostrand, 1863); J. G. Barnard, *The Peninsular Campaign and Its Antecedents, as Developed by the Report of Maj.-Gen. Geo. B. McClellan, and Other Published Documents* (New York: Van Nostrand, 1864).

14. ARCE 1865, 927–60; USWD, *The War of the Rebellion: A Compilation of the Official Records of the Union and Confederate Armies,* 70 vols. in 128 books (Washington, D.C.: GPO, 1881–1900), vol. 46; Henry Pleasants to Assistant Adjutant General IX Corps, 2 Aug. 1864, in ARCE 1864, 40–42; Andrew Atkinson Humphreys, *The Virginia Campaign of '64 and '65: The Army of the Potomac and the*

Army of the James (New York: Scribner, 1883); Henry H. Humphreys, *Andrew Atkinson Humphreys: A. Biography* (Philadelphia: John C. Winston, 1924); Richard J. Sommers, "The Dutch Gap Affair: Military Atrocities and Rights of Negro Soldiers," *CWH* 21 (March 1975): 51–64.

15. George B. Abdill, *Civil War Railroads* (Seattle: Superior, 1961); Robert C. Black III, *The Railroads of the Confederacy* (Chapel Hill: University of North Carolina Press, 1952); Angus James Johnston II, *Virginia Railroads in the Civil War* (Chapel Hill: University of North Carolina Press, 1961); Robert C. Black III, "Railroads in the Confederacy," *CWH* 7 (Sept. 1961): 231–38; James F. Doster, "Were the Southern Railroads Destroyed by the Civil War?" ibid., 310–20; Robert Bruce Sylvester, "The U.S. Military Railroad and the Siege of Petersburg," ibid., 10 (Sept. 1964): 309–16.

16. Bruce Catton, *Never Call Retreat* (Garden City, N.Y.: Doubleday, 1965), 122–32, 217–26; Quincy A. Gillmore, "The Army before Charleston in 1863," *Battles and Leaders of the Civil War,* 4 vols. (New York: Century, 1884–88), 4:58–60.

⊔⊓⊔⊓⊔⊓⊔⊓⊔⊓⊔ 5 ⊔⊓⊔⊓⊔⊓⊔⊓⊔⊓⊔

1. Moore, "National Security," 128.

2. Reports of Experiments upon the Use of Iron in the Construction of Permanent Defenses, RBE, 1866–1920, ROCE; ARCE 1866, H. Ex. Doc. 1, 39 Cong., 2 sess., pt. 3, 2:415–16.

3. Moore, "National Security," 129.

4. Reports of Experiments upon the Use of Iron in Permanent Defenses, and Report on Permanent Fortifications of the United States, RBE, ROCE; ARSecWar 1866, H. Ex. Doc. 1, 39 Cong., 2 sess., pt. 3, 2:6–7; ARCE 1867, H. Ex. Doc. 1, 40 Cong., 2 sess., pt. 2, 2:2; ARCE 1868, H. Ex. Doc. 1, 40 Cong., 3 sess., pt. 2, 2:4–6; ARCE 1869, H. Ex. Doc. 1, 41 Cong., 2 sess., pt. 2, 2:4–6.

5. Lewis, 66–72; Eugene Griffin, *Our Sea-Coast Defences* (New York: Putnam, 1885); Moore, "National Security, 129.

6. Reports on Torpedo Defenses, 1874, and ARBE 1874, RBE, ROCE; ARCE 1870, H. Ex. Doc. 1, 41 Cong., 3 sess., pt. 2, 2:4–11; ARCE 1871, H. Ex. Doc. 1, 42 Cong., 2 sess., pt. 2, 2:4–6; ARCE 1872, H. Ex. Doc. 1, 42 Cong., 3 sess., pt. 2, 2:2–3; ARCE 1873, H. Ex. Doc. 1, 43 Cong., 1 sess., pt. 2, 2:25; ARCE 1874, H. Ex. Doc. 1, 43 Cong., 2 sess., pt. 2, vol. 2, pt. 1, pp. 4–5.

7. ARCE 1875, H. Ex. Doc. 1, 44 Cong., 1 sess., pt. 2, vol. 2, pt. 1, pp. 4–5; Humphreys to Henry B. Banning, Committee on Military Affairs, House of Representatives, 11 Feb. 1876, in House Committee on Military Affairs, *Reduction of Army Officers' Pay, Reorganization of the Army, and Transfer of the Indian Bureau,* H. Rept. 354, 44 Cong., 1 sess. (1876), 170–81; Humphreys's testimony reprinted in ARCE 1876, H. Ex. Doc. 1, 44 Cong., 2 sess., pt. 2, vol. 2, pt. 1, pp. 4–6, 30; ARCE 1877, H. Ex. Doc. 1, 45 Cong., 2 sess., pt. 2, vol. 2, pt. 1, p. 4; ARCE 1878, H. Ex. Doc. 1, 45 Cong., 3 sess., pt. 2, vol. 1, pt. 1, pp. 4–6, 28–30; ARCE 1879, H. Ex. Doc. 1, 46 Cong., 2 sess., pt. 2, vol. 2, pt. 1, pp. 4–5; Henry L. Abbott, Report upon Experiments and Investigations to Develop a System of Submarine Mines for Defending the Harbors of the United States, 1880, RBE, ROCE; Moore, "National Security," 129. Documents on fort designs are in Correspondence, Blueprints, and Reports Relating to Defense, 1873–1918, Records Relating to Various Subjects, ROCE.

8. ARCE 1880, H. Ex. Doc. 1, 46 Cong., 3 sess., pt. 2, vol. 2, pt. 1, pp. 4–18.

9. ARCE 1884, H. Ex. Doc. 1, 48 Cong., 2 sess., pt. 2, vol. 2, pt. 1, pp. 4–10; ARCE 1881, H. Ex. Doc. 1, 47 Cong., 1 sess., pt. 2, vol. 2, pt. 1, pp. 4–14; ARCE 1882, H. Ex. Doc. 1, 47 Cong., 2 sess., pt. 2, vol. 2, pt. 1, pp. 4–8; ARCE 1883, H. Ex. Doc. 1, 48 Cong., 1 sess., pt. 2, vol. 2, pt. 1, pp. 4–5. Lieutenant Arthur L. Wagner of the 6th Infantry supported Newton's position in "The Military Necessities of the United States and the Best Provisions for Meeting Them," *JMSIUS* 5 (1884): 237–71. See also Eugene Griffin, "Our Sea-Coast Defenses," ibid., 7 (1886): 405–6; John G. D. Knight, "The Attack and Defense of Modern Fortifications, and the Latest Experience and Principles in Modern Sieges," ibid., 8 (1887): 381–404.

10. ARCE 1885, H. Ex. Doc. 1, 49 Cong., 1 sess., pt. 2, vol. 2, pt. 1, pp. 4–5; ARCE 1886, H. Ex. Doc. 1, 49 Cong., 2 sess., pt. 2, vol. 2, pt. 1, pp. 4–5; ARCE 1887, H. Ex. Doc. 1, 50 Cong., 1 sess., pt. 2, vol. 2, pt. 1, pp. 4–5; ARCE 1888, H. Ex. Doc. 1, 50 Cong., 2 sess., pt. 2, vol. 2, pt. 1, pp. 4–5; ARBE 1888, RBE, ROCE.

11. ARCE 1889, H. Ex. Doc. 1, 51 Cong., 1 sess., pt. 2, vol. 2, pt. 1, pp. 4–8; William R. King, "Guns and Forts," *JMSIUS* 13 (1892): 1055–75.

12. ARIG 1893, 4:715, and ARComGen 1893, 1:62–63, H. Ex. Doc. 1, 53 Cong., 2 sess., pt. 2.

13. ARBE 1894–99, RBE, ROCE; ARCE 1894, H. Ex. Doc. 1, 53 Cong., 3 sess., pt. 2, vol. 2, pt. 1, p. 12; ARCE 1895, H. Doc. 2, 54 Cong., 1 sess., vol. 2, pt. 1, p. 13; ARCE 1896, H. Doc. 2, 54 Cong., 2 sess., vol. 2, pt. 1, p. 8; ARCE 1897, H. Doc. 2, 55 Cong., 2 sess., vol. 3, pt. 1, pp. 8–9; ARCE 1899, H. Doc. 2, 56 Cong., 1 sess., vol. 4, pt. 1, pp. 13–14.

14. John S. Billings, *A Report on Barracks and Hospitals with Descriptions of Military Posts,* Circular No. 4, Surgeon General's Office (Washington, D.C.: GPO, 1870); USWD, Surgeon General's Office (John S. Billings), *A Report on the Hygiene of the United States Army, with Descriptions of Military Posts,* Circular No. 8, Surgeon General's Office (Washington, D.C.: GPO, 1875); ARSecWar 1875, H. Ex. Doc. 1, 44 Cong., 1 sess., 6; Clary, *These Relics of Barbarism;* Clary, *Life Which Is Gregarious.*

15. Thomas M. Anderson, "Army Posts, Barracks, and Quarters," *JMSIUS* 1 (1881): 446.

16. ARIG 1889, H. Ex. Doc. 1, 51 Cong., 1 sess., pt. 2, vol. 1, p. 132; ARIG 1890, H. Ex. Doc. 1, 51 Cong., 2 sess., pt. 2, vol. 1, pp. 100–101. ARSurGen every year retailed the conditions in casemate quarters with great and critical exactitude.

17. ARQMG 1884, H. Ex. Doc. 1, 48 Cong., 2 sess., pt. 2, vol. 1, p. 313; Colonel George W. Getty to Quartermaster General, 17 June 1881, in ARQMG 1881, H. Ex. Doc. 1, 47 Cong., 2 sess., pt. 2, vol. 1, p. 449.

18. ARSecWar 1902, H. Doc. 2, 57 Cong., 2 sess., vol. 1, pp. 22–23.

19. RO at Fort Monroe, 1866–68, Letters Sent, Records of Fort Monroe, ROCE; ARCE 1866, 426–27; ARCE 1867, 12; ARCE 1868, 15–16. For Brewerton, see Heitman, 1:243.

20. RO at Fort Monroe, 1869, Letters Sent, Records of Fort Monroe, ROCE; ARCE 1869, 15–16. For Craighill, see Heitman, 1:334.

21. RO at Fort Monroe, 1870–79, Letters Sent, Records of Fort Monroe, ROCE; Plans for Batteries for Heavy Guns at Fort Monroe, Correspondence, Blueprints and Reports Relating to Defense, ROCE; ARCE 1870, 22; ARCE 1871, 18; ARCE 1872, 15; ARCE 1873, 16–17, 25–26; ARCE 1874, pt. 1, pp. 19–20, 29; ARCE 1875, pt. 1, pp. 19–20; ARCE 1876, pt. 1, p. 21; ARCE 1877, pt. 1, p. 17; ARCE 1878, pt. 1, p. 20; ARCE 1879, pt. 1, p. 24. For Gillmore, see Heitman, 1:457–58.

22.   Plan for Batteries at Fort Monroe, 1880, Correspondence, Blueprints, and Reports Relating to Defense, ROCE; RO at Fort Monroe, 1880, Letters Sent, Records of Fort Monroe, ROCE; ARCE 1880, pt. 1, p. 40.

23.   RO at Fort Monroe, 1880–88, Letters Sent, Records of Fort Monroe, ROCE; ARCE 1880, pt. 1, pp. 40–41; ARCE 1881, pt. 1, pp. 39–40; ARCE 1882, pt. 1, pp. 36–37; ARCE 1883, pt. 1, pp. 32–34; ARCE 1884, pt. 1, pp. 37–39; ARCE 1885, pt. 1, pp. 31–32; ARCE 1886, pt. 1, pp. 30–32; ARCE 1888, pt. 1, pp. 3, 8.

24.   RO at Fort Monroe, 1888–90, Letters Sent, Records of Fort Monroe, ROCE; ARCE 1888, pt. 1, p. 326, pt. 2, pp. 804–6; ARCE 1889, pt. 1, pp. 12, 463–66; ARCE 1890, H. Ex. Doc. 1, 51 Cong., 2 sess., pt. 2, vol. 2, pt. 1, pp. 9–10, 385–87. For Hains, see Heitman, 1:487.

25.   ARCE 1890, pt. 1, pp. 5–7.

26.   RO at Fort Monroe, 1891–97, Letters Sent, Records of Fort Monroe, ROCE; ARCE 1891, H. Ex. Doc. 1, 52 Cong., 1 sess., pt. 2, vol. 2, pt. 1, pp. 10–11, 529–31; ARCE 1892, H. Ex. Doc. 1, 52 Cong., 2 sess., pt. 2, vol. 2, pt. 1, pp. 14–15; ARCE 1893, H. Ex. Doc. 1, 53 Cong., 2 sess., pt. 2, vol. 2, pt. 1, pp. 13–14, 635–45; ARCE 1894, pt. 1, p. 14; ARCE 1895, pt. 1, pp. 9–10, 511–15; ARCE 1896, pt. 1, pp. 16–17, 495–501; ARCE 1897, pt. 1, p. 16.

27.   RO at Fort Monroe, 1895–96, Letters Sent, Records of Fort Monroe, ROCE; ARCE 1895, pt. 1, pp. 13, 508–11; ARCE 1896, pt. 1, pp. 7–8, 16–17, 495–501. For Zinn and Casey, see Heitman, 1:1069, 289.

28.   Casey Annual Report on Fortifications, Letters Sent, Records of Fort Monroe, ROCE, and in ARCE 1898, H. Doc. 2, 55 Cong., 3 sess., vol. 4, pt. 1, pp. 687–88; ARCE 1900, H. Doc. 2, 56 Cong., 2 sess., vol. 7, pt. 1, pp. 11–12; ARCE 1901, H. Doc. 2, 57 Cong., 2 sess., vol. 2, pt. 1, pp. 11–12; ARCE 1903, H. Doc. 2, 58 Cong., 2 sess., vol. 9, pt. 1, p. 13; ARCE 1904, H. Doc. 2, 58 Cong., 3 sess., vol. 5, pt. 1, p. 10; ARCE 1905, H. Doc. 2, 59 Cong., 1 sess., vol. 5, pt. 1, pp. 11, 722.

29.   ARCE 1866, 427. There are some operational records of Fort Wool in the period immediately after the Civil War, filed with Daily RO Relating to the Rebuilding of Fort Calhoun, 1861–66, Records of Fort Monroe, ROCE.

30.   RO at Fort Wool, 1866–70, Letters Sent, Records of Fort Monroe, ROCE; ARCE 1867, 12; ARCE 1868, 16; ARCE 1869, 16; ARCE 1870, 22.

31.   Report on Modification of Fort Wool, December 1870, RBE, ROCE; ARCE 1870, 22; ARCE 1871, 19; ARCE 1872, 15; ARCE 1873, pt. 1, p. 17; ARCE 1874, pt. 1, p. 20; ARCE 1875, pt. 1, p. 20; ARCE 1877, pt. 1, p. 17; ARCE 1878, pt. 1, pp. 20–21; ARCE 1879, pt. 1, p. 34; Weinert and Arthur, 147.

32.   Plan for Modification of Fort Wool, 14 June 1879, RBE, ROCE; ARCE 1879, pt. 1, pp. 24, 34; Weinert, "Saga of Old Fort Wool," 8.

38.   ARCE 1880, pt. 1, pp. 40–41; ARCE 1882, pt. 1, p. 40; ARCE 1883, pt. 1, p. 34; ARCE 1884, pt. 1, pp. 39–40; ARCE 1885, pt. 1, pp. 32–33; ARCE 1886, pt. 1, pp. 32–33; ARCE 1898, pt. 1, p. 688; ARCE 1901, pt. 1, p. 812.

⎍⎍⎍⎍⎍⎍⎍⎍⎍⎍ **6** ⎍⎍⎍⎍⎍⎍⎍⎍⎍⎍

1.   ARComGen 1880, H. Ex. Doc. 1, 46 Cong., 3 sess., pt. 2, vol. 1, pp. 5–6; ARComGen 1881, H. Ex. Doc. 1, 47 Cong., 1 sess., pt. 2, vol. 1, p. 34. See also Jack D. Foner, *The United States Soldier between Two Wars: Army Life and Reforms, 1865–*

1898 (New York: Humanities Press, 1970); John S. Goff, *Robert Todd Lincoln: A Man in His Own Right* (Norman: University of Oklahoma Press, 1969); Richard Allen Andrews, "Years of Frustration: William T. Sherman, the Army, and Reform, 1869–1883" (Ph.D. diss., Northwestern University, 1968); Donna Marie Thomas, "Army Reform in America: The Crucial Years, 1876–1881" (Ph.D. diss., University of Florida, 1981).

2. Lewis, 75–77; William E. Birkheimer, *Historical Sketch of the Organization, Administration, Material, and Tactics of the Artillery, United States Army* (1884; rept. New York: Greenwood, 1968).See also Henry L. Abbott, *Course of Lectures upon the Defense of the Sea-Coast of the United States* (New York: Van Nostrand, 1888); Fairfax Downey, *Sound of the Guns: The Story of American Artillery from the Ancient and Honorable Company to the Atom Cannon and Guided Missile* (New York: McKay, 1956); Albert Manucy, *Artillery through the Ages* (Washington, D.C.: GPO, 1949); Harold L. Peterson, *Round Shot and Rammers* (New York: Bonanza, 1969); Richard P. Weinert, Jr., "So the Coast Artillery Is Gone—But Not Forgotten," *PJCAMP* 10 (Fall 1978): 20–31; Eugene M. Emme, "Technological Change and Western Military Thought," *MA* 24 (Winter 1960–61): 6–19.

3. ARSecWar 1883, H. Ex. Doc. 1, 48 Cong., 1 sess., pt. 2, vol. 1, p. 16; 16 Stat. 474; Edward Ranson, "The Endicott Board of 1885–86 and the Coast Defenses," *MA* 31 (Summer 1967): 74–75; ARComGen 1884, H. Ex. Doc. 1, 48 Cong., 1 sess., pt. 2, vol. 1, p. 49; Paul A. Hutton, *Phil Sheridan and His Army* (Lincoln: University of Nebraska Press, 1985), 352–53.

4. ARCE 1884, pt. 1, pp. 9–10; ARBE 1884, RBE, ROCE (also in ARCE 1884, pt. 1, pp. 55–60).

5. G. N. Whistler, "The Artillery Organization of the Future," *JMSIUS* 5 (1884): 324–30.

6. Wagner, "Military Necessities of the United States," 240, 246.

7. Ranson, "Endicott Board," 74; 23 Stat. 434. See also Moore, "National Security," 129–30; Rowena A. Reed, "The Endicott Board: Vision and Reality," *PJCAMP* 11 (Summer 1979): 3–17.

8. *Report of the Board on Fortifications or Other Defenses Appointed by the President of the United States under the Provisions of the Act of Congress Approved March 3, 1885* (the Endicott Report), H. Ex. Doc. 49, 49 Cong., 1 sess. (1886). It went to the Senate with *Letter from the Secretary of War, Transmitting Report of the Board on Fortifications and Other Defenses*, S. Ex. Doc. 43, 49 Cong., 1 sess. (1886). The Endicott Report also appeared in ARCE 1886, pt. 1, and for convenience all page references cited here are to that publication. The Endicott Board's minutes and other records are in Records of the Board on Fortifications or Other Defenses, 1885–87, ROCE.

9. Endicott Report, ARCE 1886, pt. 1, pp. 499, 500.

10. Ibid., 499–525.

11. Ibid., 515, 523, 524–25; Ranson, "Endicott Board," 77.

12. Ranson, "Endicott Board," 79; Lewis, *Seacoast Fortifications,* 77–89.

13. ARCE 1887, pt. 1, p. 8.

14. Peter S. Michie, "The Personnel of Sea-Coast Defense," *JMSIUS* 8 (1887): 1–17; ARSecWar 1887, H. Ex. Doc. 1, 50 Cong., 1 sess., pt. 2, vol. 1, pp. 23, 28; Moore, "National Security," 130; ARSecWar 1890, H. Ex. Doc. 1, 51 Cong., 2 sess., pt. 2, vol. 1, p. 5.

15.   ARCE 1890, pt. 1, pp. 4–5.

16.   See especially the following articles in *JMSIUS:* John Gibbon, "The Danger to the Country from the Lack of Preparation for War," 11 (1890): 16–28; David D. Porter, "Fortifications and Fleets," ibid., 147–58; James Chester, "The Position-Finding and Position-Designating Service in the Sea-Coast Defenses," 13 (1892): 227–48; Henry L. Abbott, "Coast Defense, Including Submarine Mines," 15 (1894): 451–67; Peter C. Hains, "Should the Fixed Coast Defenses of the United States Be Transferred to the Navy?" ibid., 233–56; George W. Van Deusen, "Which Are More Needed for Our Future Protection, More War-Ships or Better Coast Defenses?" ibid., 986–93.

17.   ARSecWar 1891, H. Ex. Doc. 1, 52 Cong., 1 sess., pt. 2, vol. 1, pp. 3, 4; ARCE 1892, pt. 1, pp. 8–9.

18.   Ranson, "Endicott Board," 77–78; ARCE 1895, pt. 1, pp. 5–6.

19.   ARCE 1896, pt. 1, pp. 8–9, 10–11, 21.

20.   Moore, "National Security," 130; ARCE 1897, pt. 1, p. 9; ARCE 1898, pt. 1, p. 11; ARCE 1899, pt. 1, pp. 9–14. For an engineer's view of problems in the early Endicott works, see E. Eveleth Winslow, *Notes on Seacoast Fortification,* Occasional Papers No. 61, U.S. Army Engineer School (Washington, D.C.: GPO, 1920).

21.   Some of these developments are discussed in Clary, *Life Which Is Gregarious,* 108–26.

22.   Ranson, "Endicott Board," 78; Graham A. Cosmas, "From Order to Chaos: The War Department, the National Guard, and Military Policy, 1898," *MA* 29 (Fall 1965): 105–6; ARCE 1898, pt. 1, pp. 8–9, 12–13; ARCE 1899, pt. 1, pp. 9–13; ARCE 1900, pt. 1, pp. 8, 11.

23.   ARCE 1901, pt. 1, pp. 7, 10–11; Moore, "National Security," 130; ARIG 1902, H. Doc. 2, 57 Cong., 2 sess., vol. 1, p. 405; ARCE 1903, pt. 1, pp. 10–12; ARCE 1905, pt. 1, pp. 6, 8.

24.   ARCE 1880, pt. 1, p. 40; ARBE 1884, RBE, ROCE (ARCE 1884, pt. 1, pp. 10, 62–64).

25.   Endicott Report, ARCE 1886, pt. 1, pp. 502–3, 510, 515–19.

26.   ARCE 1886, pt. 1, p. 5; BE to Chief of Engineers, 2 Aug. 1887, RBE, ROCE; ARCE 1887, pt. 1, pp. 10–11; ARCE 1888, pt. 1, p. 6.

27.   Proceedings of the BE, 1889 and 1890, RBE, ROCE; ARCE 1889, pt. 1, pp. 5–8; ARCE 1890, pt. 1, p. 7.

28.   Proceedings of the Board, 1890 and 1891, RBE, ROCE; ARCE 1891, pt. 1, pp. 4–6, 8, 12–13; ARCE 1892, pt. 1, p. 8.

29.   RO at Fort Monroe, 1891–92, Records of Fort Monroe, ROCE; ARCE 1891, pt. 1, p. 8; ARCE 1892, pt. 1, pp. 4, 8; George A. Zinn, "Demolition of Concrete Gun-Platforms and Magazines at Fort Monroe, Va.," *JUSA* 1 (Oct. 1892): 392–96.

30.   ARCE 1893, pt. 1, pp. 9, 16; ARCE 1894, pt. 1, pp. 5–6, 10; ARCE 1895, pt. 1, p. 9; ARCE 1896, pt. 1, pp. 16, 492–95; ARCE 1897, pt. 1, p. 16; RO at Fort Monroe, 1893–97, Records of Fort Monroe, ROCE; Weinert and Arthur, 166.

31.   RO at Fort Monroe, 1896–98, Records of Fort Monroe, ROCE; ARCE 1896, pt. 1, pp. 494–95; ARCE 1897, pt. 1, pp. 16, 656–70; ARCE 1898, pt. 1, p. 23; Weinert and Arthur, 166–67.

32.   Abbott, "Coast Defense, Including Submarine Mines," 461; ARCE 1896, pt. 1, p. 16.

33. RO at Fort Monroe, 1897–99, and RO Relating to Fortifications Constructed Separately under National Defense Appropriations, March 1898–March 1899, Records of Fort Monroe, ROCE; ARCE 1897, pt. 1, pp. 16, 661–62; ARCE 1898, pt. 1, pp. 23–24, 679–82; ARCE 1899, pt. 1, pp. 834–45; ARCE 1900, pt. 1, pp. 23, 25; Weinert and Arthur, 167–68.

34. RO at Fort Monroe, 1899–1904, Records of Fort Monroe, ROCE; ARCE 1899, pt. 1, pp. 23–25; ARCE 1900, pt. 1, pp. 23–24, 886–97; ARCE 1901, pt. 1, pp. 802–12; ARCE 1902, pt. 1, pp. 23–24, 710–20; ARCE 1904, pt. 4, pp. 3721–22; ARCE 1905, pt. 3, pp. 3009–11; Weinert and Arthur, 173–74.

35. RO at Fort Monroe, 1901–5, Records of Fort Monroe, ROCE; ARCE 1901, pt. 1, pp. 12–13; ARCE 1902, pt. 1, p. 13; ARCE 1904, pt. 4, pp. 1322 (plans for datum points), 3721; ARCE 1905, pt. 3, pp. 3009–11 (design for waterproofing). For Winslow, see Heitman, 1:1050.

36. Reports of Superintendent James Ware Relating to Fort Wool, 1905–8, and RO at Fort Wool, 1904–7, Records of Fort Monroe, ROCE; ARCE 1902, pt. 1, pp. 710–20l; Weinert and Arthur, 174; Weinert, "Saga of Old Fort Wool," 8–10; Beard, "Castle of Rip Raps."

37. During the largest target practice ever held at Fort Monroe, on 21 July 1910, a 12-inch rifle in Battery DeRussy blew up, killing eleven men and starting fires. Not just treasure but blood had been spent in something that proved to be nothing but show. The Endicott works never faced a real test. Weinert and Arthur, 178.

38. Moore, "National Security," 130; Lewis, 98–100. See also H. L. Hawthorne, "The Role of the Navy in Sea-Coast Defense," *JMSIUS* 23 (1898): 392–408; Beach, *United States Navy,* 330–34, 424–43. A prolific author, Mahan is best remembered today for his classic, *The Influence of Sea Power upon History, 1660–1783* (Boston: Little, Brown, 1890, 1918).

39. ARIG 1899, H. Doc. 2, 56 Cong., 1 sess., vol. 1, pp. 102–3; ARSecWar 1903, H. Doc. 2, 57 Cong., 2 sess., vol. 1, pp. 8–10.

40. ARCE 1902, pt. 1, pp. 6–9.

41. Ibid., 8, 11; ARCE 1903, pt. 1, pp. 8–9; ARCE 1904, pt. 1, pp. 5–6.

42. Senate Committee on Military Affairs, *Coast Defenses of the United States and the Insular Possessions* (the Taft Report), S. Doc. 248, 59 Cong., 1 sess. (1906); Lewis, 89–94; ARCE 1906, H. Doc. 2, 59 Cong., 2 sess., 5. An example of current thinking after the Taft Report is E. E. Winslow, *Lectures on Seacoast Defense* (Washington, D.C.: U.S. Army Engineer School Press, 1909). See also Paul D. Bunker, "The Mine Defense of Harbors: Its History, Principles, Relation to Other Elements of Defense, and Tactical Employment," *JUSA* 41 (March–April 1914): 129–70.

43. ARCE 1906, 5–11; ARCE 1907, 3 pts. (Washington, D.C.: GPO, 1907), 5–12; ARCE 1908, 3 pts. (Washington, D.C.: GPO, 1908), 10; ARCE 1909, 3 pts. (Washington, D.C.: GPO, 1909), 10–23; ARCE 1910, 3 pts. (Washington, D.C.: GPO, 1910), 12–24; ARCE 1911, 3 pts. (Washington, D.C.: GPO, 1911), 7–19; ARCE 1912, 3 pts. (Washington, D.C.: GPO, 1912), 7–17; ARCE 1913, 3 pts. (Washington, D.C.: GPO, 1913), 6–14; ARCE 1914, 3 pts. (Washington, D.C.: GPO, 1914), 6–15.

44. ARCE 1892, pt. 1, pp. 10–11.

45. ARCE 1900, pt. 1, p. 6; ARCE 1901, pt. 1, p. 6; ARCE 1906, pt. 1, p. 5.

46. ARCE 1909, 17; House Committee on Appropriations, *The Fortifications Appropriations Bill (Fiscal Year 1914): Hearings before the Subcommittee,* 62 Cong., 3 sess. (Washington, D.C.: GPO, 1912), 6–9.

47. *Fortifications Appropriations Bill (Fiscal Year 1914)*, 26, 45–51; ARCE 1913, 9; ARCE 1914, 9.

48. ARCE 1915, 3 pts. (Washington, D.C.: GPO, 1915), 7; ARCE 1916, 3 pts. (Washington, D.C.: GPO, 1916), 6–7; ARCE 1917, 3 pts. (Washington, D.C.: GPO, 1917), 8; Weinert and Arthur, 184.

49. Weinert and Arthur, 183–97.

50. Moore, "National Security," 127.

51. ARCE 1915, 6–16; ARCE 1916, 6–17.

52. ARCE 1917, 8–20; ARCE 1918, 3 pts. (Washington, D.C.: GPO, 1918), 34–50; ARCE 1919, 3 pts. (Washington, D.C.: GPO, 1919), 67–77.

53. ARCE 1920, 3 pts. (Washington, D.C.: GPO, 1920), 31–35. See Winslow, *Notes on Seacoast Fortification*, 438, for E. Eveleth Winslow's assertion in 1920 that justification for the coast defenses lay in the fact that the United States had not been attacked since they were built—an inverted *post hoc, ergo propter hoc* argument that persists in more recent nuclear deterrence thinking.

54. Lewis, 100–101. See also Benedict Crowell, *America's Munitions* (Washington, D.C.: GPO, 1919).

7

1. See David A. Clary, "The Biggest Regiment in the Army," *JFH* 22 (Oct. 1978): 182–84; Fine and Remington, *Corps of Engineers*, 32–41; Weigley, 395–400; Elias Huzar, *The Purse and the Sword: Control of the Army by Congress through Military Appropriations* (Ithaca, N.Y.: Cornell University Press, 1950).

2. General Order No. 4, OCE, 10 Oct. 1932, 228–10 Installation Historical File, ND; "Fort Monroe Damaged by Hurricane," *CAJ* 76 (Nov.–Dec. 1933): 440–41; G. R. Young, "The Fort Monroe Seawall," *ME* 28 (July–August 1936): 261–65.

3. Fine and Remington, *Corps of Engineers*, 52.

4. Lewis, 101–2; Weinert and Arthur, 205–6; William W. Hibbs to Chief, Fortifications Section, ND, 30 June 1942, 228–10 Installation Historical File, ND. See also Weigley, 412–13; Alfred F. Hurley, *Billy Mitchell: Crusader for Air Power* (New York: Watts, 1964); Thomas H. Greer, *The Development of Doctrine in the Army Air Arm, 1917–1941* (Montgomery, Ala.: Air University, 1955); Alfred Goldberg, ed., *A History of the United States Air Force, 1907–1957* (Princeton, N.J.: Van Nostrand, 1957); Constance McLaughlin Green, Harry C. Thomson, and Peter C. Roots, *The Ordnance Department: Planning Munitions for War,* United States Army in World War II: The Technical Services (Washington, D.C.: DA, 1955).

5. Lewis, 103–10; Fine and Remington, *Corps of Engineers*, 44n; Stetson Conn, Rose C. Engelman, and Byron Fairchild, *Guarding the United States and Its Outposts,* United States Army in World War II: The Western Hemisphere (Washington, D.C.: DA, 1964), 49–50.

6. Weinert and Arthur, 207, 228–30.

7. Lewis, 110–15; D. P. Kirchner and E. R. Lewis, "American Harbor Defenses: The Final Era," *USNIP* 94 (Jan. 1968): 84–98. The Japanese mounted 18-inch guns on two battleships in the 1940s, but they apparently would not have matched the American 16-inch if there had been a test between them.

8. ARCE 1921, 2 pts. (Washington, D.C.: GPO, 1921), 54–65; ARCE 1922, 2 pts. (Washington, D.C.: GPO, 1922), 42–50; "The Coast Artillery School Centennial Exercises," *CAJ* 60 (June 1924): 443–51; Newport News *Daily Press,* 24 May 1924.

9. Wertenbaker, *Norfolk,* 361–62; Weinert and Arthur, 207, 223; "Notes from Fort Monroe," *CAJ* 76 (July-Aug. 1933): 304.

10. Conn et al., *Guarding the United States,* 46.

11. Lewis, 115–16; Conn et al., *Guarding the United States,* 46–47; Kirchner and Lewis, "American Harbor Defenses," 98.

12. Quoted in Conn et al., *Guarding the United States,* 47.

13. Ibid., 48; John Whiteclay Chambers II, *The North Atlantic Engineers: A History of the North Atlantic Division and Its Predecessors in the U.S. Army Corps of Engineers, 1775–1975* (New York: Corps of Engineers, 1980), 64–65; Lewis, *Seacoast Fortifications,* 116–18, 122. See also G. M. Barnes, *Weapons of World War II* (New York: Van Nostrand, 1947).

14. Richard P. Weinert, Jr., "Fort John Custis and Chesapeake Defenses," *PJCAMP* 14 (March 1986): 31–36; Weinert and Arthur, 219–25; Conn et al., *Guarding the United States,* 47–49, 53; Franklin W. Reese, "Third Coast Artillery District," *CAJ* 84 (1941): 284–85, 381–82, 494, 630; Franklin W. Reese, "Fort Monroe," ibid., 85 (1942): 82–84.

15. Conn et al., *Guarding the United States,* 53–54; Weinert and Arthur, 220, 226–27, 232; Weinert, "Saga of Old Fort Wool," 10.

16. Weinert and Arthur, 234–35.

17. Weinert, "Saga of Old Fort Wool," 10–12.

18. Conn et al., *Guarding the United States,* 53–54; Weinert and Arthur, 220, 246; Lewis, *Seacoast Fortifications,* 119; Alonzo F. Colonna, "Harbor Defenses of Chesapeake Bay," *CAJ* 88 (Sept.–Oct. 1945): 84, and 89 (Jan.–Feb. 1946): 88; "Military Construction, War Years, 1940–1946: A Narrative Account," n.d., 228–10 Installation Historical File, ND. See also Otto L. Nelson, *National Security and the General Staff* (Washington, D.C.: Infantry Journal Press, 1946).

19. Weinert and Arthur, 249–52; Lewis, 124–25; Chambers, *North Atlantic Engineers,* 65.

20. Charles F. Marsh, *The Hampton Roads Communities in World War II* (Chapel Hill: University of North Carolina Press, 1950), 3. See also R. Elberton Smith, *The Army and Economic Mobilization,* United States Army in World War II: The War Department (Washington, D.C.: DA, 1959).

21. ARCE 1941, 2 pts. (Washington, D.C.: GPO, 1941), 5; ARCE 1942, 2 pts. (Washington, D.C.: GPO, 1942), 4; Fine and Remington, *Corps of Engineers,* 440–76; Eugene Reybold, *Engineers in World War II: A Tribute* (Fort Belvoir, Va.: Corps of Engineers, 1945), 3.

22. George F. Wigger to Division Engineer, Middle Atlantic Division, 4 Dec. 1943, and to Finance Officer, 2 March 1944, and Raymond H. Kelsey to Dan L. Burdette, 2 Nov. 1943, 228–10 Installation Historical File, ND.

23. "Military Construction, War Years"; Fine and Remington, *Corps of Engineers,* 187, 312–13, 327, 610; Delmar S. Lenzner, "The New Submarine Mine Depot," *CAJ* 83 (1940): 537–38.

24. "Military Construction, War Years"; Joe Jones, *1–B Soldier* (New York: Harper and Brothers, 1944), 33–35.

25. "Military Construction, War Years"; Fine and Remington, *Corps of Engineers,* 559–61.

26. Fine and Remington, *Corps of Engineers,* 447, 618–19, 621, 624–26, 635–37.

27. David A. Clary, *A History of the Army Engineers in the Norfolk District, 1607–1984* (Bloomington, Ind.: David A. Clary and Associates, 1984).

Bibliography of Primary Sources

Archives and Collections

The Records of the Office of the Chief of Engineers, Record Group 77 of the National Archives and Records Service, are the principal source of documentation on the history of the Corps of Engineers. Photographs are located in the Still Pictures Branch of the National Archives, while drawings are located in Alexandria, Virginia, at the National Archives Center for Cartographic and Architectural Archives. A useful collection of materials is National Archives Microcopy 417, Buell Collection of Historical Documents Relating to the Corps of Engineers, 1801–19, Record Group 77.

At the corps's Norfolk District Office, the 228-10 Installation Historical File, located in the Records Holding Area in Fort Norfolk, is an annually updated compilation of copies of documents pertinent to the history of the Corps of Engineers in Hampton Roads, extending back several decades.

The Government Documents and Publications Department of the Indiana University Library, Bloomington, Indiana, one of the larger and older collections of its kind, was a mainstay of the entire project. In the Office of the Chief of Engineers, Fort Belvoir, Virginia, the collections of the Historical Division are useful. The Lilly Library and Rare Books and Manuscripts Repository, Bloomington, Indiana, has some material pertinent to the Corps of Engineers, including the papers of Jonathan Williams. And no investigator of the history of the United States Army should fail to take advantage of the magnificent collections of the United States Army Military History Institute, Carlisle Barracks, Pennsylvania. Elsewhere, the Casemate Museum of Fort Monroe offers an excellent collection of materials on the history of Forts Monroe and Wool, Hampton Roads, and Chesapeake Bay. Military museums and collections around Hampton Roads include the U.S. Army Transportation Museum at Fort Eustis, the War Memorial Museum of Virginia at Newport News, and the MacArthur Memorial in Norfolk.

Printed Documents

Additional Fortifications, and an Increase of the Army. Military Affairs Document 124, 13 Cong., 1 sess. (1813). *American State Papers, Military Affairs,* 1.

American State Papers, Class V: *Military Affairs.* 7 vols., Washington, D.C.: By order of Congress, 1822–37.

Amounts of the Appropriations for Constructing and Repairing Fortifications on the Harbors and Coasts of the United States from 1815 to

1829, and of the Number of Troops Garrisoning the Same. Military Affairs Document 443, 21 Cong., 1 sess. (1830). *American State Papers, Military Affairs,* 4.

Annual Reports of the Secretary of War, Chief of Engineers, Commanding General (title varies), Inspector General (when presented), Surgeon General, Ordnance Bureau (title varies), and others, 1822–present. These were all printed before 1920 along with the Annual Report of the Secretary of War, after which combined publication ceased. From 1822, when they began, through 1837, they were published in the *American State Papers,* Class V, *Military Affairs* (7 vols.); thereafter they may be found in the Congressional Documents Serial Set, in both the Senate and House Documents. The Annual Report of the Chief of Engineers began separate publication in 1867, and continued so thereafter, although it also began to appear as a separate item in the Serial Set in 1906. Research for this study used the combined publication until that date, separate publication thereafter. Each report receives complete citation to the version used at its first citation in the text.

Armament of Fortifications. Military Affairs Document 242, 17 Cong., 2 sess. (1823). *American State Papers, Military Affairs,* 2.

Army Register, 1813. Military Affairs Document 125, 13 Cong., 1 sess. (1813). *American State Papers, Military Affairs,* 1.

Benton, Thomas Hart. *Abridgement of the Debates of Congress, from 1789 to 1856.* 16 vols. New York: Appleton, 1857–61.

Billings, John S. *A Report on Barracks and Hospitals with Descriptions of Military Posts.* Circular No. 4, Surgeon General's Office. Washington, D.C.: GPO, 1870.

Condition of the Military Establishment and Fortifications. Military Affairs Document 235, 17 Cong., 2 sess. (1822). *American State Papers, Military Affairs,* 2.

Condition of the Military Establishment and the Fortifications. Military Affairs Document 262, 18 Cong., 2 sess. (1824). *American State Papers, Military Affairs,* 2.

Condition of the Military Establishment and the Fortifications, and Returns of the Militia. Military Affairs Document 247, 18 Cong., 1 sess. (1823). *American State Papers, Military Affairs,* 2.

Delafield, Colonel R. *Report on the Art of War in Europe in 1854, 1855, and 1856.* House Executive Document (unnumbered), 36 Cong., 2 sess. (1861). Washington, D.C.: GPO, 1861.

Explanations of the Estimates for Fortifications for the Year 1837. Military Affairs Document 706, 24 Cong., 2 sess. (1836). *American State Papers, Military Affairs,* 6.

Explanatory Estimates for Rivers and Harbors, for the Military Academy, for Fortifications, and for Certain Roads during the Year 1837. Military Affairs Document 703, 24 Cong., 2 sess. (1836). *American State Papers, Military Affairs,* 6.

Fortifications. Military Affairs Document 45, 7 Cong., 1 sess. (1801). *American State Papers, Military Affairs,* 1.

Fortifications. Military Affairs Document 60, 9 Cong., 1 sess. (1806). *American State Papers, Military Affairs,* 1.

Fortifications. Military Affairs Document 84, 10 Cong., 2 sess. (1809). *American State Papers, Military Affairs,* 1.

Fortifications. Military Affairs Document 183, 16 Cong., 1 sess. (1820). *American State Papers, Military Affairs,* 2.

Fortifications. Military Affairs Document 218, 17 Cong., 1 sess. (1822). *American State Papers, Military Affairs,* 2.

Fortifications. Military Affairs Document 255, 18 Cong., 1 sess. (1824). *American State Papers, Military Affairs,* 2.

General Statement of Desertions and Deaths for Three Years Ending September 30, 1825. Military Affairs Document 326, 19 Cong., 1 sess. (1826). *American State Papers, Military Affairs,* 3.

Harris, Kenneth E., and Steven D. Tilley, comps. *Index: Journals of the Continental Congress, 1774–1789.* Washington, D.C.: National Archives and Records Service, 1976.

Heitman, Francis B. *Historical Register and Dictionary of the United States Army, from Its Organization, September 29, 1789, to March 2, 1903.* 2 vols. 1903; rept. Urbana: University of Illinois Press, 1965.

Increase of the Number and Pay of the Corps of Engineers. Military Affairs Document 418, 21 Cong., 1 sess. (1830). *American State Papers, Military Affairs,* 4.

Lands and Buildings Acquired for Military Purposes. Military Affairs Document 203, 16 Cong., 2 sess. (1821). *American State Papers, Military Affairs,* 2.

Letter from the Secretary of War, Transmitting Report of the Board on Fortifications and Other Defenses. Senate Executive Document 43, 49 Cong., 1 sess. (1886).

Letter of the Secretary of War, Transmitting Report on the Organization of the Army of the Potomac, and of Its Campaigns in Virginia and Maryland, under the Command of Maj. Gen. George B. McClellan, from July 26, 1861, to November 7, 1862. House Executive Document 15, 38 Cong., 1 sess. (1864).

Military Academy. Military Affairs Document 40, 6 Cong., 1 sess. (1800). *American State Papers, Military Affairs,* 1.

Military Academy, and Reorganization of the Army. Military Affairs Document 39, 6 Cong., 1 sess. (1800). *American State Papers, Military Affairs,* 1.

Military Force, the Posts at Which Stationed, and the Expenses of Fortifications, Arsenals, Armories, and Magazines, in the Years 1803 and 1804. Military Affairs Document 55, 8 Cong., 2 sess. (1805). *American State Papers, Military Affairs,* 1.

The Number and Cost of Fortifications, Arsenals, and Armories Completed and in Progress. Military Affairs Document 701, 24 Cong., 2 sess. (1836). *American State Papers, Military Affairs,* 6.

On an Increase of the Number of Officers and a Reorganization of ti Corps of Topographical Engineers of the Army. Military Affairs Doc ment 621, 24 Cong., 1 sess. (1836). *American State Papers, Milita Affairs,* 6.

On Making Provision for the Erection of Fortifications Proposed by tl Secretary of War for the Protection of Harbors on the Seacoast. Mil tary Affairs Document 620, 24 Cong., 1 sess. (1836). *American Sta Papers, Military Affairs,* 6.

On the Importance of the Topographical Engineers of the Army. Militar Affairs Document 465, 21 Cong., 2 sess. (1831). *American State Pc pers, Military Affairs,* 4.

On the Means Necessary for the Military and Naval Defences of the Cour try. Military Affairs Document 671, 24 Cong., 1 sess. (1836). *America State Papers, Military Affairs,* 6.

On the Necessity of an Increase of the Engineer Corps and Topographica Engineers Exclusively for Military Purposes. Military Affairs Docu ment 463, 21 Cong., 2 sess. (1831). *American State Papers, Militar Affairs,* 4.

On the Number of Cannon Required to Arm the Fortifications, the Numbe That Can Be Supplied in a Year, and on the Expediency of Establish ing a National Foundery. Military Affairs Document 680, 24 Cong., : sess. (1836). *American State Papers, Military Affairs,* 6.

On the Organization of the Topographical Engineers of the Army. Militar Affairs Document 339, 19 Cong., 2 sess. (1827). *American State Pa pers, Military Affairs,* 3.

On the Subject of an Augmentation of the Corps of Engineers of the Army Military Affairs Document 327, 19 Cong., 1 sess. (1826). *American State Papers, Military Affairs,* 3.

Organization and Allowances of the Topographical Engineers of the Army. Military Affairs Document 502, 22 Cong., 1 sess. (1832). *American State Papers, Military Affairs,* 4.

Recommendation for an Increase of the Corps of Engineers. Military Affairs Document 623, 24 Cong., 1 sess. (1836). *American State Papers, Military Affairs,* 6.

Recommendation of Appropriations for the Construction and Armament of Fortifications for the National Defence. Military Affairs Document 626, 24 Cong., 1 sess. (1836). *American State Papers, Military Affairs,* 6.

Reduction of the Army. Military Affairs Document 197, 16 Cong., 2 sess. (1820). *American State Papers, Military Affairs,* 2.

Reduction of the Army Considered. Military Affairs Document 168, 15 Cong., 2 sess. (1818). *American State Papers, Military Affairs,* 1.

Reorganization of the Army. Military Affairs Document 35, 5 Cong., 3 sess. (1798). *American State Papers, Military Affairs,* 1.

Report of the Board on Fortifications or Other Defenses Appointed by the President of the United States under the Provisions of the Act of

Congress Approved March 3, 1885. House Executive Document 49, 49
Cong., 1 sess. (1886).

Revised Report of the Board of Engineers on the Defence of the Seaboard.
In *On the Subject of an Augmentation of the Corps of Engineers of the
Army.* Military Affairs Document 327, 19 Cong., 1 sess. (1826). *Ameri-
can State Papers, Military Affairs, 3.*

*Statement of the Maximum Amount That Can Be Expended Annually
upon the Construction of Fortifications and the Ordnance Department.* Military Affairs Document 674, 24 Cong., 1 sess. (1836). *American State Papers, Military Affairs, 6.*

*Statements of the Fortifications, Their Garrisons, Steam Batteries, Etc.,
Necessary for the Defence of the Coasts of the United States, Their
Costs, Etc.* Military Affairs Document 650, 24 Cong., 1 sess. (1836).
American State Papers, Military Affairs, 6.

System of Fortifications Recommended by the Board of Engineers. Military Affairs Document 316, 19 Cong., 1 sess. (1826). *American State
Papers, Military Affairs, 3.*

Thian, Raphael P. *Legislative History of the General Staff of the Army of
the United States (Its Organization, Duties, Pay, and Allowances),
from 1775 to 1901.* Senate Document 229, 56 Cong., 2 sess. (1901).

——. *Notes Illustrating the Military Geography of the United States,
1813–1880.* 1881; rept. with addenda, Austin: University of Texas
Press, 1979.

United States. *Journals of the Continental Congress, 1774–1789.* 34 vols.
Washington, D.C.: GPO, 1934–37.

——. *Revised Statutes.* 1st ed., statutes in force 1 Dec. 1873.

——. *Revised Statutes.* 2d ed., 1878, supplements to 1901.

——. *Statutes at Large.*

——. Congress. House of Representatives. Committee on Accounts and
Expenditures of the War Department. *Expenditures of the War De-
partment.* Military Affairs Document 231, 17 Cong., 1 sess. (1822).
American State Papers, Military Affairs, 2.

——. ——. ——. Committee on Appropriations. *The Fortifications Appropriation Bill (Fiscal Year 1914): Hearings before the Subcommittee.*
62 Cong., 3 sess. (1912). Washington, D.C.: GPO, 1912.

——. ——. ——. Committee on Defence. *Fortifications.* Military Affairs
Document 43, 6 Cong., 1 sess. (1800). *American State Papers, Military Affairs, 1.*

——. ——. ——. ——. *Fortifications.* Military Affairs Document 106, 12
Cong., 1 sess. (1811). *American State Papers, Military Affairs, 1.*

——. ——. ——. ——. *Fortifications and Gunboats.* Military Affairs Document 66, 9 Cong., 2 sess. (1807). *American State Papers, Military
Affairs, 1.*

——. ——. ——. ——. *Fortifications and Gunboats.* Military Affairs Document 72, 10 Cong., 1 sess. (1807). *American State Papers, Military
Affairs, 1.*

——. ——. ——. Committee on Fortifications. *Fortifications.* Military Affairs Document 13, 3 Cong., 1 sess. (1794). *American State Papers, Military Affairs*, 1.

——. ——. ——. ——. *Fortifications*. Military Affairs Document 19, 3 Cong., 2 sess. (1794). *American State Papers, Military Affairs*, 1.

——. ——. ——. ——. *Fortifications*. Military Affairs Document 24, 3 Cong., 2 sess. (1795). *American State Papers, Military Affairs*, 1.

——. ——. ——. ——. *Fortifications*. Military Affairs Document 28, 4 Cong., 1 sess. (1796). *American State Papers, Military Affairs*, 1.

——. ——. ——. ——. *Fortifications*. Military Affairs Document 29, 4 Cong., 2 sess. (1797). *American State Papers, Military Affairs*, 1.

——. ——. ——. ——. *Fortifications*. Military Affairs Document 31, 5 Cong., 1 sess. (1797). *American State Papers, Military Affairs*, 1.

——. ——. ——. ——. *Fortifications*. Military Affairs Document 32, 5 Cong., 2 sess. (1798). *American State Papers, Military Affairs*, 1.

——. ——. ——. ——. *Fortifications, Munitions, and Increase of the Army*. Military Affairs Document 33, 5 Cong., 2 sess. (1798). *American State Papers, Military Affairs*, 1.

——. ——. ——. Committee on Military Affairs. *Army Organization*. House Report 33, 40 Cong., 3 sess. (1869).

——. ——. ——. ——. *On an Increase of the Corps of Engineers and a Reorganization of the Topographical Engineers*. Military Affairs Document 292, 19 Cong., 1 sess. (1826). *American State Papers, Military Affairs*, 3.

——. ——. ——. ——. *Reduction of Army Officer's Pay, Reorganization of the Army, and Transfer of the Indian Bureau*. House Report 354, 44 Cong., 1 sess. (1876).

——. ——. ——. Special Committee on Contract with Elijah Mix. *Contract for Stone at the Rip Raps and Old Point Comfort*. Military Affairs Document 234, 17 Cong., 1 sess. (1822). *American State Papers, Military Affairs*, 2.

——. ——. Senate. *Fortifications and Gunboats*. Military Affairs Document 64, 9 Cong., 2 sess. (1806). *American State Papers, Military Affairs*, 1.

——. ——. ——. *Fortifications and Gunboats*. Military Affairs Document 74, 10 Cong., 1 sess. (1807). *American State Papers, Military Affairs*, 1.

——. ——. ——. Committee on Military Affairs. *The Army of the United States: Its Components, Its Arms, Services, and Bureaus, Its Military and Nonmilitary Activities*. Senate Document 91, 76 Cong., 1 sess. (1939).

——. ——. ——. ——. *Coast Defenses of the United States and the Insular Possessions*. Senate Document 248, 59 Cong., 1 sess. (1906).

——. ——. ——. ——. *On Increasing the Number of Officers of the Corps of Engineers of the Army*. Military Affairs Document 588, 23 Cong., 2 sess. (1834). *American State Papers, Military Affairs*, 5.

——. ——. ——. ——. *Relative to a Gradual Increase of the Corps of Engineers and Topographical Engineers of the Army*. Military Affairs

Document 281, 18 Cong., 2 sess. (1825). *American State Papers, Military Affairs*, 3.

——. ——. ——. ——. *Report on the Means of National Defence*. Senate Executive Document 85, 28 Cong., 2 sess. (1845).

——. War Department. *Contracts Made in the Year 1818*. Military Affairs Document 175, 15 Cong., 2 sess. (1819). *American State Papers, Military Affairs*, 1.

——. ——. *Estimates for the Year 1819*. Military Affairs Document 169, 15 Cong., 2 sess. (1818). *American State Papers, Military Affairs*, 1.

——. ——. *Fortifications*. Military Affairs Document 22, 3 Cong., 2 sess. (1794). *American State Papers, Military Affairs*, 1.

——. ——. *Fortifications*. Military Affairs Document 76, 10 Cong., 1 sess. (1807). *American State Papers, Military Affairs*, 1.

——. ——. *Fortifications*. Military Affairs Document 89, 11 Cong., 2 sess. (1809). *American State Papers, Military Affairs*, 1.

——. ——. *Fortifications*. Military Affairs Document 206, 16 Cong., 2 sess. (1821). *American State Papers, Military Affairs*, 2.

——. ——. *State of the Fortifications of the United States*. Military Affairs Document 26, 4 Cong., 1 sess. (1796). *American State Papers, Military Affairs*, 1.

——. ——. *Subject Index to the General Orders and Circulars of the War Department and the Headquarters of the Army, Adjutant General's Office, from January 1, 1881, to December 31, 1911*. Washington, D.C.: GPO, 1912.

——. ——. *The War of the Rebellion: A Compilation of the Official Records of the Union and Confederate Armies*. 70 vols. in 128 books. Washington, D.C.: GPO, 1881–1900.

——. ——. *The War of the Rebellion: A Compilation of the Official Records of the Union and Confederate Armies. General Index and Additions and Corrections*. Washington, D.C.: GPO, 1901.

——. ——. Army Service Forces. *Annual Report for the Fiscal Year 1944*. Washington, D.C.: Army Service Forces, [1944].

——. ——. ——. *Annual Report for the Fiscal Year 1945*. Washington, D.C.: Army Service Forces, [1945].

——. ——. Artillery Board. *System of Nomenclature of Parts of Modern Batteries*. Engineer Mimeograph No. 60, published in Artillery Notes, No. 13. Fort Monroe: Artillery School Press, 1903.

——. ——. Corps of Engineers. *Supplement to Mimeograph No. 60, Subject: Revision of Nomenclature Relating to Modern Seacoast Batteries, December 22, 1908*. Mimeo. Washington, D.C.: Office of the Chief of Engineers, 1908.

——. ——. Quartermaster General's Office. *An Exhibit Presenting a Pictorial History of the Housing of the Army*. [Washington, D.C.: Construction Service, Quartermaster Corps, 1928.]

——. ——. ——. *Outline Description of U.S. Military Posts and Stations in the Year 1871*. Washington, D.C.: GPO, 1872.

——. ——. Surgeon General's Office [John S. Billings]. *A Report on the*

Hygiene of the United States Army, with Descriptions of Military Posts. Circular No. 8, 1 May 1875. Washington, D.C.: GPO, 1875.

Virginia, State of. *Calendar of Virginia State Papers and Other Manuscripts.* Ed. William P. Patmer, Sherwin McRae, and Henry W. Flournoy. 11 vols. Richmond: Superintendent of Public Printing, 1876–93.

——. Virginia State Ports Authority. *Annual Report, 1981–82.* Richmond, 1982.

——. ——. *Measuring the Impact of the Waterborne Commerce of the Ports of Virginia on Employment, Wages, and Other Key Indices of the Virginia Economy, 1953–1965.* Charlottesville, Va.: Bureau of Population and Economic Research, 1967.

Index